P9-DZZ-922

Common Ground

"Beautiful. It's refreshing to read a piece of place-writing that digs so deeply and tenderly into a marginal landscape, and which (strikingly) does so using a novelist's tools as well as a nature writer's."

Will Atkins, author of *The Moor*

"Boldly imaginative. . . . This is writing of the highest order. . . . Deep, rich and alive in prose that bubbles up from the fetid loam and, as in the work of, say, John Clare, William Wordsworth or DH Lawrence, takes pleasure in every micro-scopic, fecund detail . . . What makes this book special is the author's total immersion in his subject. . . . [Cowen] is an outlier, a Northern voice, a set of eyes on the soil, and *Common Ground* is his outstanding addition to—and expansion of—the canon."

Ben Myers, *Caught by the River*

"A brilliant, unusual and brave investigation of the landscape and ourselves. . . . Truly a wonderful, powerful book, written with such openness and love that I can't help feeling that—if people the world over read it—it would surely make the world a better place."

Tom Cox, author of *The Good, The Bad, The Furry*

"Formally brave, beautifully written; an intimate account of one of life's great turning points, and a luminous, painstaking exploration of place."

Melissa Harrison, author of *At Hawthorn Time*

"Heartfelt, deep, beautiful and moving."

Tristan Gooley, author of *The Natural Navigator*

"Luminous . . . A breath of fresh air."

Irish Times

"Blending natural history with a novelistic approach, Cowen revives his connection to the evocative, mysterious power of the natural world."

Sunday Express

"A poetic memoir, unearthing the transformative power of a patch of ground. This apparently scrappy and overlooked piece of wasteland—a tangle of wood, meadow, field and river—proves to be, under Cowen's forensic and magnifying gaze, brimming with riches."

Northern Echo

"Very beautiful indeed. . . . [Cowen] has all the alliterative grace and fresh metaphors of a poet."

New Books Magazine

Common Ground

*Encounters with Nature at
the Edges of Life*

WITH A NEW PREFACE

Rob Cowen

THE UNIVERSITY OF CHICAGO PRESS • *Chicago*

The University of Chicago Press, Chicago 60637
Printed in the United States of America
26 25 24 23 22 21 20 19 18 17   1 2 3 4 5

First published by Hutchinson in 2015
First published in paperback by Windmill Books in 2016

Extracts from "Swifts" taken from *Ted Hughes: Collected Poems for Children*
© Estate of Ted Hughes and reprinted by permission of Faber and Faber Ltd.

"Never Any Good" by Martin Simpson, quoted with kind permission
from Martin Simpson

ISBN-13: 978-0-226-42426-2 (cloth)
ISBN-13: 978-0-226-42443-9 (e-book)
DOI: 10.7208/chicago/9780226424439.001.0001

LIBRARY OF CONGRESS CATALOGING-IN-PUBLICATION DATA
Names: Cowen, Rob, author.
Title: Common ground : encounters with nature at
the edges of life / Rob Cowen ; with a new preface.
Description: Chicago : The University of Chicago
Press, 2016. | Includes bibliographical references.
Identifiers: LCCN 2016014736 | ISBN
9780226424262 (cloth : alk. paper) | ISBN
9780226424439 (e-book)
Subjects: LCSH: Cowen, Rob—Homes and haunts—
England—North Yorkshire. | Natural history—Eng-
land—North Yorkshire. | Human beings—Effect of
environment on—England—North Yorkshire.
Classification: LCC QH138.N845 C69 2016 | DDC
577.428/4—dc23 LC record available at
https://lccn.loc.gov/2016014736

♾ This paper meets the requirements of ANSI/NISO Z39.48–1992
(Permanence of Paper).

For my father and my son

Contents

VIADUCT

THE NIDD

BILTON BECK

THE SEWAGE WORKS

THE OLD RAILWAY

BILTON

WEIR

BILTON
HALL

THE HOLLOWAY

THE LANE

LANE

N
W E
S

NOT TO SCALE

Preface to the US Edition

Last week I was back in the edge-land, hiding in a tree line, waiting for the deer. Being midwinter in northern England the whole day had felt dull and blunt; then, as a long exhale of commuter traffic sighed through the trees, it began to darken too. The edges of late afternoon trembled; brown and black limbs of tree and hedge blurred in the quick-fading light. I could feel the chill of the frozen bank on which I was lying seep through my jacket, into skin, into bone. But if the deer were to materialise before the curtain of night fell completely I knew it would be now, into the shifting dusk. Unhooking my elbows from brambles, I lifted my binoculars and scanned the land.

When you come to know any place well, the mind gets comfortable. It can miss things. Watching for wildlife through binoculars brings a shift in perspective like going from looking at the surface of a river to thrusting your head underwater and opening your eyes. The scene came alive as it fractured into constituent parts. I paid attention to each detail zooming up in my vision, seeing them afresh: a clump of earth tufted with weeds; a sugar beet keeled over like an abandoned spinning top; plastic bags shredded in the hedges; a frequency of mist; a crow head-bobbing over the soil. At the far end of the field I could make out a rusted digger part overgrown by shrub and thorn. Then, as I watched, there was a movement above its collapsing cab. A buzzard spilled out of an ash tree and circled the field, mewing, its silhouette slipping across the gridlines of pylons and powerlines that criss-crossed the beaten-bronze sky behind.

Since writing this book I have been asked many times to define what I mean by 'edge-land'. It occurs to me now that had I been able to capture that moment it would serve as a neat reply—a shorthand sketch of the longer story you're about to begin. The way the bird's fan-tailed form was framed against sunset and infrastructure, the white noise and lights of rush-hour traffic filtered through ancient wood—all hinted at the kinds of encounters, collisions and complexities you find in the strange borderlands where human and nature intermesh. Things happen here all the time, if you're alive to them.

It was environmentalist and geographer Marion Shoard who first mooted the term 'edge-land' in 2002. She recognised that, in a geographical sense, edge-lands are challenging to pin down, calling them 'apparently unplanned, certainly uncelebrated and largely incomprehensible territory'. And she's right. These are terrains—I'm always more tempted to call them *dimensions*—that are, by nature, both evasive and universal. Evasive because they are always moving, shaped and re-shaped by the tides of change, gnawed at by the sprawl, devoured without any of the fanfare or public fuss. They are the inglorious fallow patches you find at the fringes of the everyday, the places we glimpse but overlook, the shape-shifting interzones: one moment there, the next vanished under some new development, only to emerge somewhere else as the tide of progress washes on. Yet edge-lands are also universal across the globe, found wherever we are: in the liminal spaces between urban and rural environments, the built-up and the agricultural, the civilized and the wild. They are those messy, transitory pieces of ground found at the city limits, on the edge of town and suburb; the overgrown parks, strips of old wasteland, abandoned woodlands or fields; the dark, mysterious lands surrounding a housing project or highway, shopping mall, retail complex or disused factory; the unloved tangles of scrub and high-grown vegetation fringing waterworks, sub-stations, cemeteries, railways, rivers, parking lots, turnpikes and junkyards. Each is freighted with the histories of what went before, but although traces of these things may remain eerily in their fabric, edge-lands seem to fall (temporarily at least) outside the capitalist definition of useful—i.e. worked, profitable, exploited. As such, they give the sense of being beyond the normal rules and constrictions of our otherwise ordered, commercialised and increasingly privatised world; they exist in a kind of suspended animation, neither one thing nor another. An antithesis. Running wild behind advertising hoardings, they're both ghosts of the past and apocalyptic visions of the future, haunting our self-obsessed present.

Despite their universality, this isn't a book concerned with finding and documenting edge-lands across the globe; rather, it is about the obsessive revisiting of one patch of common ground and how a forgotten wasteland on the edge of town in the north of England revealed itself to be a prism through which the wider negotiations of our lives could be seen, a place where I could truly measure the common ground that exists between human and nature, the present and the past, and each other. This is about my own transformative journey into a particular border-land and the lives and layers I found there. Nevertheless, a side effect of writing it has undoubtedly been a heightened sensitivity and inquisitiveness towards similar spaces the world over.

Today, wherever and whenever I travel, I find I'm auto-matically drawn to the margins before anywhere else, as though they somehow deliver a deeper truth than that which you find on the surface. It is that same truth Ghan-di was alluding to when he wrote: 'Whilst everything around me is ever changing, ever dying, there is underly-ing all that change a living power that is changeless, that holds all together, that creates, dissolves and recreates." Peer into the cracks in the asphalt and a junk-strewn waste ground and you can witness flashes of it manifest-ing. Sometimes it appears more dramatically. As I write this I have a browser open on my laptop with a gallery of images charting the disintegration of large tracts of New Orleans housing stock devastated by Hurricane Katrina in 2005. In what must have been an incomprehensible sce-nario to New Orleans residents before the waters rose, entire areas of the Lower Ninth Ward lie abandoned to the encroaching green. Houses have been consumed by vegetation; collapsed ceilings show nature's colonisation within. In various shots a fence has been hauled down by thorns, wildflowers erupt from a front porch and a roof is unrecognisable under a tousled mane of creeper.

After the hurricane, the very footprint of the city was altered; suburb turned to edge-land as nature erupted from the sidelines, living and breathing, beautiful and terrifying. Indeed, it wasn't long after the waters began receding that national news bulletins were rife with panicked reports that wildlife was 'taking over' the beleaguered city. There were incidents of alligators being found in swimming pools, foxes evicted from the airport, homes infested by a brown widow spider usually only found in deep forest, raccoons clearing out fishponds and even armadillos moving in to air conditioning units. Such things may come as a shock to a modern world that has distanced itself from nature, but perhaps they shouldn't. To paraphrase the Situationists of 1960s France: *Beneath the pavement, the beach.* We forget that even amid our most parcelled-out, paved-over urban environs, the other, miraculous, living world we share this planet with is never far away, and never afraid of reclaiming its turf—given half a chance.

I saw this first-hand recently when visiting the island city-state of Singapore for a literary festival. In what has to be one of the most densely built-up and contested places on the planet, it's inconceivable that any area could remain fallow for long. And it certainly appeared so as my plane landed. Singapore's waterfronts prickled with harbour-side industry and smart, new, manicured developments. Enormous skyscrapers, grand, swimming pool-topped hotels and towering financial behemoths rose from the heat-haze; bustling districts of new housing, old markets and ex-colonial institutions spread from its centre, smothering every inch of ground. Fighting jetlag, I wandered around its centre along vast boulevards flashing with brilliantly lit retail palaces, big as cathedrals. All felt accounted for and squeezed for profit and I held out little hope of recognising anything that might resemble edge-land. Yet, following a talk I gave the next day, a brilliant local botanist and university lecturer, Shawn Lum, invited me to travel a few blocks outside the CBD to experience his own local patch, a tear in the fabric that he promised was 'wild' and 'magical'. Intrigued, if a little doubtful, I rode north with him to the Buona Vista district, climbed over a road barrier and down a steep bank between two vast buildings to reach a dismantled railway. As city receded skyward, the contrast was astounding. Light, sound, smell, temperature—everything in that green gully was dizzy with life. The world closed down. We slowed down. Mimosa and cow-grass grew in a thick carpet beneath our boots as we ambled beneath crumbling old concrete bridges cracked open by the spooling coils of roots. Banana, durian, rubber and rambutan trees conjoined in deep, tangled curtains of jungle either side, resurgent echoes of the vanished villages and their fruit orchards lost under the spreading city in the late nineteenth century.

Shawn explained that the old railway was once part of a line connecting Malaysia to Singapore's harbour but, after being decommissioned, eighteen miles of track had become caught in contested ownership and were left to bloom back into chaotic life. Ahead of us birds skittered between the undergrowth and plantain squirrels jumped about the tree canopy. Stopping to drain our water supplies in the fierce humidity we listened to the reedy, grass-between-the-thumbs shriek of Asian koels—a cuckoo of crows' nests—then sheltered from a hot rain storm down among that ubiquitous plant of old railway sidings: bindweed or *Morning Glory*. Exactly the same plant cloaks the disused railway in my own edge-land in Yorkshire, albeit with one colourful difference. What is a pure white funnel flower seven thousand miles away in Britain was, in the tropics, an extraordinary purple. Nevertheless, its familiarity made me smile. To suddenly have the sense of home soil in the heart of Singapore was unexpected and grounding. Picking a flower and pressing its vivid iridescence into my notebook's pages, I felt both incredibly insignificant and profoundly connected to the earth.

Later, when high up on the fifty-seventh floor of my hotel, I looked for, but failed to locate, that green artery on any city maps. Despite some impressive-looking parks and 'nature areas' around Singapore, that wild, semi-subterranean corridor was nowhere to be seen. It runs, for now at least, unmarked. And I have a romantic notion of it running thus forever, gloriously neglected by the thrusting ambitions of the city above.

A crucial part of any edge-land's identity is this same sense of having fallen through the gaps, of being *un*managed and largely unofficial. As such, wherever they are in the world, edge-lands have the sense of being last vestiges of truly wild spaces, different to more regulated models of wilderness that you can encounter in nature reserves and national parks. In those kinds of places, the notion of 'wild' can also mean unnaturally preserved, tidied up, ecologically arrested and prescribed; they can be landscaped areas where—necessarily and for good reason—visitors are encouraged to stay on the footpath, picnic only in designated areas and strictly follow codes and signboards. Being unregulated and unpoliced, however, edge-lands provide a freer space for ecosystems to flourish unchecked. Left to their own devices they run riot with plants, insects and animals that might in other places be controlled or destroyed. Species burst forth into tatty, intricate, interwoven habitats providing plentiful sources of food and shelter, creating intense corridors of wildlife that exist cheek-by-jowl with built-up environments. To be able to walk into them each day, to be still and attend to unexpected, bright choruses of birdsong, a seething trail of ants amid the leaf litter or butterflies drifting from invasive wildflower to invasive wildflower, is to be delivered time and again into the possibility of escape; escape from the constrictions of our human world, perhaps even ourselves. Edge-lands provide portals into other worlds and other, vivid lives that exist in parallel to our own, and we need that as surely as we need anything.

In 2009, the United Nations released a report that highlighted a fundamental shift in our relationship with nature. 'By the middle of [this year],' it stated, 'the number of people living in urban areas (3.42 billion) surpassed the number living in rural areas (3.41 billion)." In what is a first for human history the report confirmed that more people now live in the urban environment than the rural. This means that—although you may not realise it—for the majority of the world's population, edge-lands are probably our closest 'natural' or 'wild' terrain. Yet in America, just as in Britain, 'natural' and 'wild' are concepts that traditionally tend to be thought of as distinctly separate from us. Nature is a place (usually) that's entirely removed from the ordinary modern world in which we live and work. It is somewhere or something grand and remote that must be prepared for and travelled to, far beyond the city and the town. Nature is a two-week vacation, a bumper sticker, a summer camp. In this scenario, edge-lands on the threshold of city or suburb are written-off as the trashy fade-out of the urban experience, somewhere that must be passed through on the way to the true majesty of grandiose wilderness. Keeping this great category divide between human and nature suits us, of course. There are other purposes to it—it helps us to keep in our minds the notion that nature is something other to our species, something visitable, controllable. It keeps at arm's length the frightening reality of that great, final wilderness where we all must find ourselves one day. But the truth is that nature isn't just the grizzly on Mt Washburn or the wolf in the pines; it's also the backyard crow and creeper; the starling and the cancer cell. It is around us and among us and—ultimately—we're all beholden to it. Nature was before us and it will be after us. Edge-lands attest to this like nowhere else. They record and replay our negotiation with it, bridging the gap and reminding us that to think of our existence as removed from the wider biosphere is nothing short of delusional. With this awareness comes incredible payback. You begin to see that you can never be alone in nature. Lonely, perhaps, but never alone. There are riches in the wastelands and they're yours to find.

As definitions go, that's hardly the most helpful, but perhaps it's the most accurate I can give. As I said—things happen here all the time, if you're alive to them.

Buzzard and bronze sky were gone by the time the deer drifted from cover. They came, though, shadows formed from shadows, framed against blue-blacking sky, ducking through the hedge to step flightily, twitching-eared, over weedy field, past the decaying digger and into the thorn bushes, where they disintegrated into dark again. Creatures in flux; energies coursing through the land and me. The thrill of sheer life. There, then gone.

Harrogate, January 2016

To know fully even one field or one land is a lifetime's experience ... a gap in a hedge, a smooth rock surfacing a narrow lane, a view of a woody meadow, the stream at the junction of four small fields – these are as much as a man can fully experience.

Patrick Kavanagh, 'The Parish and the Universe'

Prologue

NEW YEAR'S EVE

Maps transform us. They make birds of us all. They reveal the patterns of our existence and unlock our cages. If it wasn't for that map, a second-hand Ordnance Survey given as a Christmas present, maybe none of this would have happened. It was New Year's Eve and I lay on the bed with the town unfolded before me. I felt tired; constrained; racked with cabin fever. I needed to get out. From a circle of Biro drawn around my new house I flew up and over the unfamiliar rooftops and roads, past shops, schools, hair salons and bookmakers, seeking the nearest open ground. Below me suburbia slunk down a shallow hill towards an endless patchwork of delineated farmland. Hemmed in between the two, I saw it: a tract of white paper, tree symbols and the varicose vein of a river. It lured me down, eyes to paper, body to freezing earth.

Somewhere a bell struck five as I cut through the start-stop traffic of the ring road. Exhaust fumes swirled fog-like, landlocked by the plummeting temperature. Underfoot the afternoon rain was hardening into a slippery film; frost feathered lawns. That peculiar post-Christmas malaise, thick with burning coal, pressed down on the houses. As the shrivelled sun disappeared into the mass of pitched roofs, chimney stacks and telegraph wires, I flowed on past a plastic Santa on a roof with no chimney and along a trench of emerging street lights. Either side of me, rows of Victorian terraces morphed into post-war semis before, finally, modern red-brick boxes whirled off the road in car-cluttered cul-de-sacs. Then, after a mile of walking, even their low walls and privet hedges began to thin. Through the gaps the dark, dank countryside of northern England rose like a great black wave.

At the bottom of the hill a rough track bisected the road suddenly and steadfastly, tracing a contour with nineteenth-century arrogance. It was a definitive border. Light and vegetation were in accord. Dimness shrouded the land beyond. Among the bare blackthorn, ash and spider-limbed elder, I spied relics: soot-blackened sandstone walls, riveted iron plates and the overgrown ditch and mound of a siding. It all uttered a single word: *railway*. A footstep and I had crossed from the bright lights and right angles of bulbs and bricks into black bushes and trees, whose infinitesimal branches overlapped the track like hair growing over a scar. Unwittingly the railway was fulfilling a different function now – this was the high water mark of the sprawl. Suburbia washed against its southern bank in a mass of rickety fences and scattered bin bags disembowelled by brambles. Down its northern side the town dissolved into something other: a kind of wildness. Winter-beaten meadows stretched into wood before the earth rose again as field and hill that met sky in an unbroken ridge.

I hunkered down by a fence and tried to take it all in. Nothing stirred. There were hints of shapes forming in the distance – stands of larch, pylons, barns – but they were impossible to distinguish. The road I'd followed narrowed and wandered past a squat pub crouching in a hollow, then became lost in the rawness of fields. Tarmac turned to footpath, footpath into soil. Marking the border on opposite sides of the road were two vast oaks thirty metres high. Entwined above me their limbs created an arch, ancient sentinels guarding a forgotten world. I knew it, though. The urban fringe. The no man's land between town and country; this was the edge of things.

I can't say what imperceptible force drew me there, only that I needed to reach it. That frontier called me. Maybe a speck of its soil carried in a starling's foot had been drawn down deep into my respiratory system, circulating around my bloodstream and lodging on my temporal lobes, establishing itself as a point of reckoning. Whatever it was, I felt a sense of returning, like a bee to a hive. Weeks had passed since I'd left London with the weightlessness of new horizons in Yorkshire, the place I'd grown up, but far from being the liberating experience I'd imagined, moving house had proved to be an imprisonment. For too long I'd been stuck in an unending cycle of working, painting walls and unpacking boxes, sleeping fitfully in rooms that stank of gloss, acid in my throat, numbed by the cold of open windows. I'd find a whole day had slipped by as I sifted through collections of things that suddenly seemed to belong to a previous life. I'd hardly ventured into the world outside. Soon the shortening days and wintry gloom made familiarising myself with new surroundings even harder. All my routines were jumbled; every light switch was in the wrong place. In truth, the act of handing over the keys to my London flat had signified a greater shift: present to past. All the maps I had once navigated my life by – the routes to work, streets, cafés, flats, parks and pubs – were redundant. They covered a region 220 miles to the south. I was stuck somewhere else, between tenses, between spaces, between lives.

Everything changes continuously, of course, nature is perpetual flux, but we are good at suppressing uncomfortable reminders of the greater cycles. We rope ourselves to imagined, controllable permanence. Clocks are wound to the rhythms of modern anthropocentric existence: the nine-to-five grind, career trajectories, the working week, Saturday nights out, summer holidays, twenty-five-year mortgages, pension plans, retirement. It's how the adverts metronome our lives. Yet staring out over that edge rendered such things irrelevant. Time was a different animal, *in*different, a deer running unseen through the trees. There was nothing by which I might measure the moments passing until the rise and fall of a siren shrieked through town, then silence again. With the cold, clear, descending dark came euphoria; it prickled my neck and released the atom-deep sensation of otherworldliness. It was the blur of joy and terror felt when facing something prior to and greater than the self. My pulse slowed as the adrenalin dispersed and for a second I imagined it was my cells recalibrating to the deeper rhythms of the dark, my body resetting to the land.

Resetting is the right word, for I have long loved the edges. As a boy I would jump the fences around my home town to walk and play in the scrubby penumbra between the urban and rural. I remember snapshots: watching badger setts at night above lagoons of amber-lit houses; seeing a grass snake slide past my single-buckle school sandal; catching the sapphire flash of kingfishers as cars droned over a road bridge above. Those seams were wide, exotic kingdoms possessing a kind of condensed wildness precisely because of their proximity to the civilised. A feeling of being alive and in the moment used to fill me when watching the unfurling of leaves in spring or the dying back of bracken in autumn. For a boy growing up in 1980s Britain, such things provided a vital counterpoint to the increasingly fabricated reality of classrooms, kitchen tables and TV.

Not that I could have explained as much at the time. The fringes of towns were regarded as worthless, scrappy, litter-filled areas. The 1980s, hard-edged with modernity, eyes fixed on slick futures, were a triumph of ownership and boorish capitalism; the margins didn't just fall through the gaps, they were the gaps. At best they were work-shy and unrealised. At worst, dangerous. Parents warned of tetanus-carrying beer cans, used condoms and addicts' needles. If you believed what teachers told you, every pond teemed with Weil's disease and every wood held a resident paedophile. Perhaps because of the proliferation of these myths, mine was the last generation to lay claim to the edges in a meaningful sense. A great societal shift indoors was already in motion. When the first of my friends got a VHS player and TV in his bedroom, lines of us sat cross-legged in wordless rapture; it was as though God had materialised on his chest of drawers. Watching *Star Wars* for the first time was as altering and exciting to young minds as voyaging into space itself. The borders between town and country soon became even emptier, trembling in the periphery of our vision, rarely coming into focus. A seam of silver birch here, a shock of wildflowers there – dimly remembered dreams glimpsed from passing cars and trains.

Of course, edges by their very nature are always being drawn and redrawn but a relentless force shapes the urban fringe. Developers had done for the places I loved by the last years of the 1980s – about the same time I began daily journeys into a city for school. Wood and bracken field crumpled and creased under bulldozers to be replaced by stacks of executive homes and double garages. Lawns were laid and the scrubland that had hidden dens tilled into river-rock flowerbeds. It was odd to watch the deconstruction of my childhood in such literal form. My connection to the fringes altered then too; the needle of my internal compass pointed to the exciting new worlds of adolescence. Once traced, however, these perimeters are never truly lost. They are only ever picked up and flexed. There are always edge-lands for those that seek them.

A dog barked somewhere. How long had I been sitting in darkness? I tried to get my bearings. I retrieved the map zipped into my jacket pocket and, using the light of my phone's screen, found my position. The meadow I was in ran along the old railway north-west for half a mile before meeting the snaking curve of a river twisting east. The topography of the railway and river formed two sides of a harp-shaped piece of ground perhaps only half a square mile in size. Ahead I could just about make out the entrance to a narrow lane. It was little more than a hole in the trees but by my reckoning it twisted up the eastern side of this region all the way to the river, creating the final side to what looked like an inverted triangle. Seeing the space laid out like this I was gripped by the urge to explore, to align myself with somewhere that, like me, was caught between states. I pushed myself up and jumped the fence again.

The lane was claustrophobic with darkness. The road crumbled into uneven track beneath my trainers. Thin trees, somehow blacker than the inky air, pressed inwards.

I fumbled across a beck as a wash of moon illuminated the earth. I had trespassed into the passageway of a house where nobody lived. Hazel and beech formed dilapidated walls and dog prints threadbare patches in a carpet of frost. Another hundred yards and I breached into open fields where cheek-numbing air testified to hard, anthracite land to the north. Pylons disappeared west into gloom, roped to one another. I thought of climbers crossing a glacier in a blizzard. The path was little more than a silver thread stitched along a frayed blackthorn hedge, running along the length of a field before linking with an opposite thicket of hawthorn. Together they encircled the unmistakable form of a holloway. This exuded the same mysterious grav-itational pull of all ancient paths; I was drawn into its high-hedged heart, aware I must be tracing footsteps sculpted over millennia.

How does one know that a house is empty even before knocking at the door? A sensation? A mix of stimuli? We do, though. And it was the same in that tunnel of trees. Everything had the feeling of time-slipped abandonment and loss. Quarried stones running down the track spoke of centuries-old shoring up for hoofed beast and cart. My arm scraped a holly, stirring the whispers of men who first cut these paths when all England was forest. Ahead a different sound: a hiss like an unlit cooker ring leaking gas. Suddenly the banks around me grew higher as the path sunk, drop-ping me into a wooded ravine. I smelled the mercurial stench of river water and knew I'd reached the northernmost boundary.

Sure enough a sweep of river materialised. At the bottom of a gorge, through the silhouettes of trunks, its surface was a sheen of pewter stained with the patina of reflected trees. The hiss came from where it slid over a wide cobbled weir in a sheet of foam. The break was brief; within a few feet it returned to reflection and I walked westward in a similar state, upstream, into the black wood. I saw only one other colour: a blue fertiliser bag bobbing around the roots of an alder. Even defiled by litter, the river seemed to possess an intractable power – a seam of continuousness cutting through deep, silent strata. Night rendered the trees both three-dimensional and completely flat, as if a cubist painting. Knuckled roots poking through rotting leaves became branches at chest height; distant trees suddenly scratched my face and hands. Then, behind me, a fox tore down the curtain of wildness with a scream. I felt my body shiver under my jacket, my breath blowing in quick, thick clouds. By the time I had scrambled back up a high, bramble bank and into open meadow again, I was lightheaded, ecstatic and bleeding. My mind flitted between disorientation and elation as I walked back over brittle grass in something like a state of bliss. Beyond the veil of singed-whiskered willowherb the geometry of the town encrusted the hill. Half-dim, half-bright pulses in the dark, orange-red, white and yellow, blurred by the cold night air. Sporadically, fireworks whined, plumed and burst over the roofs – and it all looked different, as if by walking through the edge-land I already had new eyes.

I was yet to know how powerfully this colloquial tract would come to affect me over the coming year, how intertwined with my existence and consciousness it would become; how profoundly it would alter me. The ground and its inhabitants were still to stir and take possession. But this encounter was my first experience of that patch of earth as a place of transformation. It had triggered within me a fascination that grew into obsession. Despite being in the shadow of thousands of houses, this place felt different, unclaimed, hidden.

For many years I had sought and written about the wildness encountered in the more expected places: the rarefied national park, the desolate moor, the distant mountaintop, the sweeping coast, but I'd forgotten that there is something deeper about the blurry space surrounding us all where human and nature meet. One word stayed with me: *layers*. Even before I'd started the process of investigating it in any depth I was aware that this edge-land was a crossing point where countless histories lay buried. There were its human narratives, the records of our long tangling with land – colonisation, hunting, farming, war, industry and urbanisation – but these were only part of the story. Enmeshed in every urban edge is also the continuous narrative of the subsistence of nature, pragmatic and prosaic, the million things that survive and even thrive in the fringes. This little patch of common ground was precisely that: *common*. And all the richer for it.

I began to walk through it at different times of day and night and from different directions. Some days I'd stay until there was no light left; others I'd wake up in darkness, disoriented, unsure where I was, with the haunting calls of tawny owls thrilling my ears. Other voices from the fields, woods and meadows brushed up against my consciousness, catching on my skin like the threads of spider silk. There is a depth that comes from revisiting a place relentlessly. I would pass a fallen pine and suddenly see it as a sapling breaking through the mud; I would know the ranks of people who'd sat beside it and the innumerable dogs that had cocked their legs to spray its rough scales. I would see the river not as a man but as a mayfly. I'd approach hares with the tread of a medieval trapper. Tracing the screaming arcs of swifts, I could feel thermals above as keenly as they did. I began to perceive the stories of everything that stepped, slid and swooped over my patch of common ground, to see through an increasing array of eyes and know myriad existences. And at the same time, the land, its layers and inhabitants seemed to be ever more bound up in things happening in my own life.

I started carrying a notebook with me everywhere, filling its pages with my experiences. Some days I'd drop off in the grass and dream of changing shape. It wasn't madness, it was a growing awareness born of watching and it brought with it relief and a flood of understanding: if I could dissolve myself from the human shackles of logic and reason, I thought, I might achieve an immersion, perhaps a written expression *in common* that goes beyond the sterility of the field guide, the dry social history or the overblown romantic eulogy, something drawn from observation, heightened awareness and sensitivity; something akin to a truer shape of the place I'd found.

Some years ago, by an arc of stone and another river, I stood outside the sealed entrance to the Chauvet cave, halfway up a cliff face in the Ardèche Gorges. Lost for millennia but rediscovered in the 1960s, it was found to contain over 400 representations dating from between 15,000 and 30,000 years ago. The mind can struggle to take in that reach of time, especially when you consider the pyramids are 'only' 5,000 years old. The handprints and multiple forms painted, scratched and worked within show the land outside the walls as it once was. *Life,* as it once was. To enter it now you need written permission from the French interior minister but its secrets can be seen and studied in the masses of high-definition photographs, film footage, interactive websites, even a full-scale, life-size and painstakingly accurate reproduction of large sections of the cave being built in a hangar nearby. Beasts massing to cross the river are represented on a section where the buttery flank of wall narrows, giving the same sense of compressing, jostling and funnelling you see with animals congregating at a water's edge. Astoundingly beautifully rendered aurochs and rhinoceros dissolve into one another; lifelike horse heads merge perspectives. Owls, bears and panthers emerge and change shape in what is a dark, frightening and fascinating world. There are strange human forms too, all of which seem at a point of transformation. On one piece of rock is a half-human, half-animal couple: the man has the right leg and arm of a human and a buffalo head. The woman ends as lioness. They embody a now inconceivable engagement between animal, land and human, a sort of becoming-animal.

In a similar way, it didn't feel strange to cross between narratives in what I was writing. I felt myself moving with a sort of Chauvet cave-like focus and freedom through the creatures, characters and stories I encountered over the passage of that year. What I didn't realise until later is that in seeking to unlock, discover and make sense of a place, I was invariably doing the same to myself. The portrait was also of me.

Once upon a time the edges were the places we knew best. They were our common ground. Times were hard and spare but the margins around homesteads, villages and towns sustained us. People grazed livestock and collected deadfall for fuel. Access and usage became enshrined as rights and recognised in law. Pigs trotted through trees during 'pannage' – the acorn season from Michaelmas to

Martinmas – certain types of game were hunted for the table and heather and fern were cut for bedding. Mushrooms, fruits and berries would be foraged and dried for winter; honey taken from wild beehives; chestnuts hoarded, ground and stored as flour. The fringes provided playgrounds for kids and illicit bedrooms for lovers. Whether consciously or not, these spaces kept us in time and rooted to the rhythms of land and nature. Feet cloyed with clay, we oriented ourselves by rain and sun, day and night, seasons, the slow spinning of stars.

Humans are creatures of habit: we all still go to edges to get perspective, to be sustained and reborn. Recreation is still re-creation after a fashion, only now it occurs in largely virtual worlds. Clouds, hyper-real TV shows, 3D films, multiplayer games, online stores and social media networks – these are today's areas of common ground, the terrains where people meet, work, hunt, play, learn, fall in love even. Ours is a world growing yet shrinking, connected yet isolated, all-knowing but without knowledge. It is one of breadth, shallowness and the endless swimming through cyberspace. All is speed and surface. Digging down deeper into an overlooked patch of ground, one that (in a global sense, at least) few people will ever know about and even fewer visit, felt like the antithesis to all of this. And it felt vitally important. You see, I still believe in the importance of edges. Lying just beyond our doors and fences, the enmeshed borders where human and nature collide are microcosms of our world at large, an extraordinary, exquisite world that is growing closer to the edge every day. These spaces reassert a vital truth: nature isn't just some remote mountain or protected park. It is all around us. It is in us. It *is* us.

CROSSING POINT

I am dreaming of the edge-land again. It has begun to colonise my sleeping mind. Dreams take place in the midst of Scots pines and down among the cold, scrub-scattered banks. I am following a fox, a copper coat floating through the trees. He pauses. A backward glance. Incredible eyes – coronal black holes over exploding suns, that intense face; mouth curled at its edges in the white, greasepaint smile of The Joker. Another step. *Am I to follow?* He pads up to the lip of a rise and disappears. Suddenly I can't move. I wake. The weak glow of a street light forms an exclamation mark on the ceiling. I dress quietly, shivering in the dark, pick up my notebook and walk out.

Modern life is such that it can be hard to see beyond the present. You think you know somewhere, but really you only know a layer, a moment. Most people don't even notice such things, but just look around you. The moss-swollen pavement crack, the rosette of a dandelion defying a driveway or a gutter-growing sow thistle, these are glimpses of what lies beneath and beyond. The deep past and the far future.

A map drawn by Ely Hargrove in 1798 in his *History of the Castle, Town and Forest of Knaresborough: with Harrogate, and Its Medicinal Waters* shows the town I call home as little more than two rows of cottages. Harrogate, as the world now knows it, doesn't yet exist. This hilly stack of roads, traffic lights and pristine flowerbeds, of imperialistic hotels, antique shops, churches and promenades, is still open land. Its cottage gardens, fields and marsh-meadows await delineation, diversion, draining and deed. The 'medicinal' wells that will soon lure legions of the aspiring middle classes to holiday here or, should they strike it lucky in the mills and mines, build vast villas on the woody escarpments to the south-west, are little more than mud-edged watering holes. Pigeons pick at salt accretions forming at their rims; only the informed aristocracy and gentry shoo them away to take the waters. The arches, domes and sweeping curves of Regency and Victorian architecture that will soon form the grand structures of 'the English Spa' lie dormant, locked in the gritstone cliffs and subterranean clay of the surrounding countryside.

Hindsight imbues the map with the feeling of land on the cusp. It has the death-stare of ground destined to be choked with high-density housing, tower blocks, supermarkets, shopping centres, warehouses and car dealerships. In a matter of decades the two little rows of cottages will bloom into an urban mass that consumes the surrounding land and villages. Eventually it will reach an ancient settlement a mile and a half to the north marked on Hargrove's map with a green blob – a legacy of its past life as part of a royal hunting forest. Between the 1950s and 1990s, the sprawl will swallow Bilton almost entirely, appropriating its Celtic name – *farmstead of 'Bilain'* – for the suburb thrown up around the scattering of old homesteads and farms. But it will be a last meal. For now at least, Harrogate will reach no further in this direction. Bilton will become edge-land. There will be no protests, no public outcries or petitions, no organised lines of conservationists standing in front of diggers or activists hauled down from centuries-old oak trees. The ground won't resist sublimation. After all, it has always been a place of transience and transformation. It has known innumerable beginnings and endings.

In contrast to the raw, jump-in-head-first shock I'd felt on the night of its discovery, my preliminary forays into this new-found land were to take more methodical lines. Confronted with this unknown world stripped bare by winter, I planned to navigate via its most obvious physical structures and landmarks in an effort to map and taxonomise it. I felt I needed to gain a sense of its definable perimeters and the logical starting point was its western edge.

In the 1840s, Britain's burgeoning railway network reached Harrogate. Or, more accurately, it reached its outskirts. A decree had been passed to prevent the town's reputation for restorative waters, clean air and new regal façades from being besmirched by steam-spewing engines and dirty tracks. Instead, it was decided that the first rail link should end a mile to the east, down a hill at a cluster of old houses named after the little stream that flowed past them. Starbeck station birthed a thriving community. Rows of terraces, pubs and hotels sprang up around the marshalling yards and engine sheds. Horse-drawn coaches more aesthetically acceptable than coal-fired trains carried the great and the good up the hill to the unsoiled spa resort. Meanwhile, financiers and speculators gripped by the frenzy of nineteenth-century Railway Mania had already turned their eyes to the land beyond, prospecting its gullies and ridges for potentially lucrative routes that would lead further north.

The intention of the hastily formed Leeds and Thirsk Railway Company was clear from its name: to connect the thrusting might of an industrialised Leeds with the outlying city of Ripon and the market towns of Thirsk and Northallerton. Starbeck soon changed from terminus to thoroughfare. Track was hammered at startling speed along a contour heading north-north-west, skirting Harrogate in a sweeping curve, colluding with natural features where possible and running over earth-stacked sidings where it wasn't. Lying on its path like a body on the tracks was Bilton.

Seated in an ornate Leeds office, no doubt with a ticking clock and glowing dog grate in the corner, a suited and bespectacled planning clerk drew a line in pencil. That was all it took. The course of the railway sliced scalpel-like through the community, straight over Bilton Lane, an old drovers' road that had already seen 400 years of foot- and hoof-fall. The bisection created an 'X' of road and rail, necessitating a level crossing. Probably no more than two white-painted wooden gates with lamps on top, it was a crossing nonetheless. X marked the spot and it still does, for today this is the edge-land's point of origin, its moment of departure from the housing estates, cul-de-sacs and crescents; it is where town becomes something else.

I'm sure such coincidences must occur frequently in the buffers between urban and rural worlds. Over time, people and landscape leave unintentional impressions on each other. Things assume significance impossible to predict or design in the moment they are conceived. Though the planning clerk is dead and the railway gone, the crossing point remains.

It was an afternoon in January and I had finished work early and returned to follow that pale seam of rubble and mud, heading north-west from where it jutted off at ninety degrees away from the divisions of tarmac, B&Q plank

fencing and houses at the end of Bilton Lane. The dismantled railway line was much as I imagined it would be with
rail, sleeper and shingle removed: its edges grew unchecked
with bramble, dog rose and willowherb. Overshadowing it
were the interwoven tangles of blackthorn, hawthorn,
willow, hazel, elder and ash that become indecipherable
when denuded by winter and silhouetted by a low sun.
Light torched the cascades of dead grass and birds flapped
between branches in shrill fly-pasts, needling the air.
Everything else was still. Poised.

A long, rectangular block of masonry and concrete,
green with algae, was almost entirely consumed by bare
vegetation. I would find out later that this was an old raised
platform constructed in the 1880s to unload the coal that
supplied Harrogate's new gas works built to the west of
Bilton at New Park. Its by-products, vats of ammonia and
bitumen, were ferried back here and loaded onto trains
heading for Middlesbrough and the shipbuilding yards
of the north-east. It's an exchange commemorated in the
bulbous liquorice residues that still dribble down
the platform's face – great black drips hanging frozen in
perpetual movement.

A few steps on and the ground gave way on either side,
giving the impression of being on an elevated causeway. To
my left, through the bare shrubbery, the siding became a
sea wall holding back waves of housing. Tidy terracotta
boxes with grey roofs rolled with the landscape's contours
like a swelling ocean, its peaks and troughs awash with the
debris of suburbia: wires, cars, caravans and, cresting
the waves, the square tower of a church, a tree or two and
a dull defiance of offices. To the north the land assumed
the form of a sloping field, dipping down to a farmhouse
and beck, then quickly gaining height and thickening with

tawny wood. Before long both sides rose again to rejoin the old railway. The land flattened out ahead, disappearing into that curious imperceptibility of distance.

Neat staves of high-tension cables ran perpendicular overhead, east to west, carried underarm by pylons. The nearest one stamped down the brush beside the track with four barbed-wire-rimmed feet and wore a thin shawl of starlings around its shoulders. The fence of power line skirted the bulbous edge of Bilton, disappearing westward and corralling the town as it fell away downhill towards a sewage works and the old hamlet of Knox. To my left as I walked, houses petered out into a sward of common land, a grassy plateau fringed with willow wood and birch copse that accompanied the old railway onwards. A hay cropping meadow that in summer would be sewn with bird's-foot trefoil, orchids, Welsh poppies and alive with the rhythms of crickets, it was scarred with the marks of the urban: wonky white goalposts rusting in damp air, shrubs and gorse bushes fruiting the odd multicoloured membranous bag of dog shit.

Mirroring it on my right, through a thin belt of vegetation, I recognised the meadow I had stumbled back over in darkness on New Year's Eve. The trace of a path cut across its dead grass and disappeared into a dark intensity of trees running parallel to the old railway, 300 yards east. Seeing it from the opposite direction elicited a similar feeling as when I'd first come here; it was like there was something undisclosed in the grass, brush and branches, something alternative. But I didn't change course. Materialising through the mist ahead was what I'd come to find – the conclusion to this western border. Amid the blur of alder and beech was a huge metal gate prickling with railings and razor wire. The old railway plunged headlong into these reinforced shutters. Walking to the side, I craned my neck, expecting to watch the track's crumbling demise down the wooded gorge into the river. Instead, an unbroken viaduct spanned the deep, narrow valley.

In 1846, having progressed north-west half a mile from the crossing point at Bilton, the Leeds and Thirsk Railway Company faced a major obstacle. The River Nidd's meandering course rises on the mountain flanks of Great Whernside in the Yorkshire Dales, flowing on to the River Ouse near York before winding a further fifty miles east to the Humber Estuary and, eventually, the North Sea. This immovable, looping line was also proving a problem for men working a few miles east laying track for the East & West Yorkshire Junction Railway trying to connect York to Starbeck. Simultaneously, both companies began the gargantuan task of bridging the river, one at Bilton, the other at Knaresborough. Teams of men quarried enormous gritstone slabs straight from the sides of the Nidd gorge at an impressive rate. Then disaster struck. Nearing its completion in 1848, Knaresborough's viaduct dramatically collapsed, sending thousands of tonnes of stone, crenellated towers and carved abutments thundering into the water. Rebuilding took another three years. Today it is an iconic sight, ranked among the county's 'best views', immortalised in local TV news credits and preserved in the digital repositories of countless visitors. Upstream, Bilton Viaduct suffered a different fate. Opening without incident in 1847, it towered 104 feet in height with seven arches that dutifully carried freight and people over the water for well over a century. The railway's closure in 1964 heralded an unglamorous downfall. Unneeded, unnoticed, I found it shut-off, shackled and destitute, left to the plumes of dead Oxford ragwort and buddleia that bristled from its cracks.

Mostly we have no idea what surrounds us. We don't care. But to me the viaduct's scale and size seemed extraordinary, so too the sense of rectitude, the way the abandoned arches reflected nobly, silently in the river. Pines and bare larch furred the far bank; wide black water flowed beneath. An irony, I suppose, but this once great crossing point was now a definitive end, closing the western border of the edge-land with steel and wire and by virtue of its sheer height above the river. There was no plaque or information board, just words scrawled over the metal shutters and mesh: 'Kurt has Hep C' in crude, white letters. The font was difficult to age, having an almost 1970s punkish, sectarian quality, like a Belfast wall, and yet the name and disease suggested a more modern story. In the dusk-darkening trees, this combination of viaduct and graffiti felt like a worn memorial to vanished narratives, fragments of time, lost lives. Here was an arbitrary bridge between the solid and the sensory. I thought of the men who built the viaduct clambering over its sides on ropes and wooden platforms, and of the water below sliding over slabs of rock that will outlive me. I thought of Kurt, his disgruntled lover and the randomness of what is lost and what is passed on. Of time passed and time passing.

Stupid, dangerous, but I wanted to get closer.

Manoeuvring gingerly around the railings, I shimmied up and onto the top of the three-metre-high shutters, smashing my feet against them with a loud bang, balancing precariously, a leg on either side. As the drop to the river below dizzied my vision I felt a moment of vertigo but pushed through it to tip my body over, dropping heavily onto the viaduct, heart pounding. There, from that lofty position above the slow, sliding river, I could see the shape of the edge-land from the other side. The breadth and depth of the landscape, past, present and future, era stacked on top of era. There lay the northern border, the ancient serpentine Nidd, twisting east on its course through flak-like explosions of trees. Westward, where the horizon vanished, hundreds of rooks and jackdaws were swarming rookeries, rattling, squeaking and murmuring in the furthest sepia crowns, jostling for position, bickering, fluttering up and settling again. They turned the bare branches black. I thought how each must be the offspring of the victorious or the lucky, a culmination of a bloodline dating back incalculable years. Out where the river gorge slumped into fields, the white and yellow orbs of street lights demarked the western rim of Harrogate. The sewage works was an abandoned city turning and whistling to itself. Bare sycamores towered over illuminated suburban avenues, stark against the ashen sky. Closer were beeches whose forms resembled milky streams of hearth smoke rising from cottages. A light, strong and gold, burned by the river's edge in a clearing. In the descending gloom it passed for a great bonfire.

Standing there on the margin, listening to the faraway chatter of swarming corvids and watching the spectacle of night drawing a veil over the river, time and space seemed to slip and reel. All at once, I was on top of the viaduct; down by the fire and among the feathery swarm of rooks. Although it had appeared to reach its end, the long, straight track of the old railway had derailed me into a multiplicity of time, body and space. The air was thick with the sour-sweet tang of slurry, leaf-litter and pine. Then a more urgent, sweaty-smoky reek clamped over my nose and mouth, that rancid but unmistakable coming alive of irrepressible earth and animal. *Fox.*

Where? I wanted to see it. I wanted to glimpse the creature I shared this timeless twilight with. Keeping out of sight, I scanned the woods from my position high in the canopy, before clambering back over the shutters to search among the brambles, finding nothing. A few days later, though, it found me.

The smell was there when I returned at dusk to carry on plotting the next side of the perimeter, the edge-land's northern boundary. Turning right at the viaduct, I took a rough track leading east along the edge of the meadow's curtain of trees and down into the wood. Despite the onset of night, I followed the Nidd downstream, guided by the weak circle of a head torch, past drowned trees and along a muddy edge. The water tricked and teased, appearing still, not even a ripple giving away movement. I noticed a branch and a plastic cider bottle held in its surface overtaking me. The sudden presence of the fox was just as bewildering. Its scent, strong and sharp as cut lemons, crowded, pressed and pushed me, as though the animal was dancing between my legs, mocking my cumbersome, slithering progress. At moments I was sure it must be right behind or beside me, but each time I turned, my beam only emphasised the wood's emptiness, silvering briefly the bars of beech and oak bristling the banks.

I fell back to the task in hand: making notes on distance travelled, cross-checking my location with the old OS map and striking a rough outline of the edge-land in my note-book. The next human mark wasn't far away: the large weir I'd found on New Year's Eve and, in my drawings at least, the region's easternmost point from where it rose up and headed south in a final continuous track all the way back to the crossing point. The weir announced its position first, roaring with the previous night's rainfall. My torchlight touched the long raised hump that spanned the river; water broke over it into a spumy waterfall that seethed down its cobbled slope. Lurking in the shadows of the far bank was a building whose small windows set in a solid, strong wall gave the impression of a scowl – the architectural expression of industry. Newly rendered stonework and posh cars parked in its drive suggested a private house, but the weir was a giveaway of its origins. Mills had once mirrored each other on opposite banks. One had crumbled, one survived; now the water held the only reflection. The ground where I stood still had the ghost of foundations, a muddy clearing pocked and chipped with roots and buried stones. Nearby, presumably set up to take advantage of the view of the Nidd's continuing journey east, I found the stone-ringed blackened ground and ash of an old campfire. It had been lit about the same place as the ruined mill's hearth would once have been. Here the smell of the fox was strongest. I rounded the trees and breathed its musty trace, trying to discern a route, stepping, pausing, sniffing and moving again until I lost the trail near an uprooted beech. Some way off to my right I heard the fleet-footed scarper of animal through brush. *It's marking territory*, I thought. *Just like me.*

One final pencil stroke needed drawing: the eastern border. It was the connecting line of the perimeter and, fittingly, the way I'd first entered the edge-land. Now it would take me back the opposite way. I headed up a set of steps and then on to a track called Milner's Lane – most likely a derivation of 'Miller's Lane' – which rises south through the woodland to where the trees funnel, telescope-like, into the holloway. In places the interwoven saplings and shrubs squeezed in so closely that they felt like hands holding my shoulders. I pushed on with the sense I was swimming for the surface from too deep in a dark sea. Red aircraft warning flares winked from turbines on far-off moors. Closer was Harrogate's nebula of lights flashing like a zoetrope through the black stems and twigs. Then I was out and breathing. Behind me, the aperture of the holloway shrivelled as I strode towards the crossing point over moonlit fields, descending through a last few hundred yards of wooded tunnel rank with the stink of fox. Over the crossing point, at the street lights of Bilton Lane, I closed my notebook and looked back. The dots were joined; I had navigated some form of the edge-land's limits. But even as I traced that final line with my pen I knew all attempts at fixity were an illusion. Scratching the surface had already revealed far more than I could hope to contain in a cartographic triangle.

I see the fox for the first time on the same day I lose my job. It is an amicable split over lunchtime beers but no less worrying. 'The business is downsizing. Last in first out. I'm very sorry.' His hands are shaking so I make it easy. Besides, there is a catch in his voice, a fear for his own future. Well-founded, as it turns out. Everywhere is talk of cuts, job losses and economic turmoil the likes of which the world has never seen. Across my own industry, in writing and journalism, fees are being dropped and positions slashed. Writing, it seems, has gone from a profession to a luxury, an aside indulged in by those who have a 'real' job too. I heard a minister talking on the radio yesterday saying: 'Given we are in a recession, anything that cannot justify its existence financially has to go.' *What, like education?* I wanted to shout. *Like libraries? Like a reed warbler? Like love?* And go where exactly? Out to the shed with a revolver?

Do the honourable thing and leave the world to those who think solely in economic terms? How did we ever get this far, confusing what is necessary for life with what living is about? To make it worse, I couldn't shake the image of him slinking off the mic, taking a congratulatory call from the party whip and letting his thoughts turn to an upcoming two-week break in Tuscany or somewhere. It seemed such a removed existence, so unreal.

I call Rosie, my wife, who is still working most of the week in London, and she falls quiet when I tell her the news. 'It'll be OK,' she says, even though we're both thinking the opposite. We bury our fears about moments like this but it's only ever a shallow grave. The profound uncertainty of it all is dredged from our unconscious as we sleep, manifesting as a strange 4 a.m. panic, a cold world that (you hope) melts away with the daylight.

After leaving the pub I forgo the quietness of our empty house for the edge-land and wander along the lane up into the fields. I need space. I'm vexed by an earworm: the words 'recession bites' – currently in every strapline and news VT – bounce around my head. It is a curious, irregular verb 'to bite', but it feels appropriate. Almost the first thing I find is a crow, an offering, headless, its chest open and the keel of its breastbone picked clean white. It lies back in the mud with its wings spread as if it has fallen out of the sky. I know this is the fox's work. I can smell it everywhere. Chances are it was disturbed and marked its prey, intending to return. Nearby I find a torn hole pushed through the vegetation, a gap in the fuzz where new growth has been restricted by regular passage. The tatty circle promises open field beyond and, crouching, I can make out a shallow trench disappearing up and over the dirt. Half-pissed and growing sick of my human skin, I push through.

A pink-grey film of sunset behind pylons reflects in puddles between the dark, ploughed peaks of soil. Shards of sky shine in a hundred tiny fjords, briefly turning earth to heaven. Pressed into the churned ground between my feet is a single deer print. Inspection reveals it is a decent size with dewclaws visible, a roebuck's. I search around but see no other, only this single hieroglyph lifted out of context months ago by a farmer's plough. Ahead, a fat woodpigeon settles and bobs into a cluster of holly; I follow and find the indentation of a deer's laying-up point scattered with white hair. Around it are clumps of badger-dug earth tilled by its bear-like paws on the hunt for a rabbit nest. It is darkening but I'm drawn further by the fox's scent, west, tracing prints, walking deep into the heart of the edge-land. Waves of brambles as high as my head catch my clothes. Each tug on jacket or trouser works free a thread and unpicks a layer in me too. Moving through this terrain it becomes impossible to hold onto the cares and concerns of town: my worries about money and work. Through the saplings and brambles, the changing light and sounds, my attention cannot rest for long on anything; all things rush up and are absorbed in a second – the dry rattle-caw of a crow concealed in a black ash, the sour reek of fox sprayed on the frigid earth.

My eyes adjust. As the fields swallow the last of the day I crawl under a hedge and trudge across soil as thick and dark as chocolate cake. There is a little wooded valley ahead and, from within it, a barking. *The fox?* No, it's more dog-like, a hoarse rasp, a smoker's cough. I slip down an escarpment to a beck where it echoes again, followed by the crash of undergrowth. A flash of white rump bounding into inky brush – *deer*. It was a roe throating a warning. I'm standing still, wondering what other creatures are moving about me in the darkness, when I realise I'm sinking. Ankle-deep sludge drowns my shoes, forcing me to wade quickly forward in blind, belching steps until, leaping a stagnant pool, I land face-first on a muddy bank.

The fox manifests as I kneel there trying to catch my breath and work out where I am. I begin to right myself when a tree's shadow morphs into an ebony silhouette, a shape from another realm trotting, head raised, along the treeline, fifteen, maybe twenty metres away. It is large, full-grown and winter-pelted, with a thick tail that it drags semi-submerged through the scrub like a rudder, scenting its wake. Seconds pass and I realise I'm holding my breath, immersed in the smell, the stillness, the sheer immediacy of it all; I'm willing it to drag me under, entranced by its indifference. *There, look*, I want to say, *there's your proof of another world, an earlier world, a greater world beyond economic justification. There tiptoes the counterpoint.* And I want to join it. Follow it. But just as quickly and quietly as it appeared, the fox slips away. The door closes. When I reach the outline of town again, the street lights stab my eyes into streams. I'm numb with cold, caked in mud, shivering with the thrill of encounter.

The red fox (*Vulpes vulpes*) has endured a long and abusive relationship with our species, much like edge-lands. However, unlike many rural mammals in the UK, which have seen a sharp decline brought about by hunting, modern farming methods and the privatisation and monoculturalism of land, the fox has adapted to whatever environment we have thrust upon it, sticking for the most part to its own set of territorial rules. This adaptability – which experts term 'biological plasticity' – has kept fox numbers steady, in spite of our attempts to destroy them and their traditional habitats. In fact, paradoxically, there has been a proliferation. They are common. Relentless. Elusive. They have forced themselves upon us, and us on them. When you look into it the figures suggest that the UK has 225,000 foxes alive at any one time, a number that almost doubles during the breeding season. As much as 14 per cent of that population is thought to inhabit our towns and cities, although given their ghostliness I'm not sure how we'd ever know for certain. Regardless, the fox is probably now more an enmeshed part of the urban experience than that of the lone walker in a field; it possesses a wildness that shocks twilight streets, a feral face that flashes across parks and patios, effortlessly slipping between worlds.

No doubt because of our begrudging coexistence over millennia, the fox has come to symbolise many things in human art and literature. As I lie curled up in bed, my skin zinging and red after a hot bath, I read of how many Far Eastern cultures once recognised this animal as the embodiment of the shape-shifter, able to move between realms and bodies, neither fully human nor animal. In other societies, the fox is a shamanistic talisman, the animal form of 'psychopomp'. I feel a twinge of excitement, for I know this word. Psychopomps are guides on spiritual journeys or rites of passage, the beings responsible for escorting souls to the afterlife. Based on this notion of the transitional creature, psychologist Carl Jung appropriated the term in the 1930s to refer to the mediator between unconscious and conscious realms. I switch out the light and listen to heavy freight rumbling a downstairs window in its frame. A low hum, like a chant. It all seems portentous, foretelling.

*

Finding the fox again is no easy business, but it feels imperative. As the last of my pay rots away and living suddenly alone, with no work to contain me, I turn away from town, beyond its jurisdiction, and spend whole days rooting through the overgrown hedges searching for tracks, kills and bone-filled twists of black fox shit. The house grows messier, but I don't mind. I'm too busy peering down the tunnel-vision perspectives of lanes over fields that look as though I'm viewing them through the wrong end of a telescope. The edge-land is overpowering at times. Consolatory, cold, late afternoons before rain are painted a beautiful duck-egg blue and pink and sweetened with drifting woodsmoke. Rooks blow across the narrow aperture of my vision like the wind-blown ash. With no foliage to subdue it, light blooms and burns in the wood, sending shadows of trees creeping down the slopes, over me, towards the river. It is as though their spirits have slipped from the trunks to drink. Along the lane or down by the Nidd, saplings caught in a rising breeze quiver and clatter in woody notes against each other. At night I listen to their xylophonic sounds in distinct octaves, their tunings dependent on each tree's thickness. They gather and crack, knock, creak, groan in communion. One second their bone beats are echoing far away, the next they are shockingly loud and right behind me, as rhythmical as falling footsteps. Walking alone in the dark, I suddenly fancy I'm being followed by a long-dead cattle drover leading his longhorns to market, a weather-cracked face under a rough hat, teeth clamped around a clay pipe. When I summon the nerve to look, he is gone; his white eyes are just holes in a holly hedge.

I find the fox beneath a pine. Actually, if I didn't know better I'd say he sprang out of the tree, one of three Scots pines that grow where the wood joins the meadow. In man's hierarchies of timber, pine is a commoner. It is the cheap stuff, the material sliced down and split for flat-pack furniture; the very name maligned as the fragrance of toilet cleaner and taxis. Compared to mahogany or walnut, the alluring cedar or the majestic oak, pine is plantation fodder, plentiful and useful but with no class. This centuries-old tree begs to differ. With its resplendent poise and beauty, it is the very essence of wildness and craggy moor, a poster boy from the ancient Caledonian Forest. Winter fiddles with colour filters dragging down tones, desaturating until the landscape takes on the drab hues of a 1940s cine film. In contrast, this tree's bright trunk glows copper, rising over a tangle of bramble to bend gracefully at the top where it is persuaded eastwards by the prevailing wind. Its lowest branches are snapped short, spiky, bristling, without bloom and with little more than a haze of ochre, but higher up the thick branches that meet its circular core are sculpted arms. Just like a waiter carrying a tray, each fawn limb holds up a nest of needles that range from silver-green to a jade green-blue. A wind stirs them into life so that the whole canopy suddenly resembles an animal rising, the mist of needles shoaling and shimmering in the way fur ripples over muscle. As the gusts build, these coalesce and then separate, kaleidoscoping the darkening steel sky behind and creating a swirling vortex of green, silver and rust. It is magnetic. A passing magpie, caught by a blast of wind under its wing, is consumed by the whirlpool and disappears into its depths. At that moment the fox is birthed below, into the meadow. At first indistinct from the reddish trunk, he trots down to the frosted grass, eyes screwed tight, blinking in the last light. When he turns back, I follow him as far as I can. True to his mythical function, the fox is escorting me into this land.

Out there my hands freeze and thaw as notes are made obsessively, messily. Rain, when it falls, blots the ink. I record where I see the fox, where he moves and what he shows me. New spidery lines are required and scribbled over my ordered maps. The rough courses of his runs are bumpily drawn, approximated, often while walking. And he is a he, I'm sure of it. Although differences between the sexes are fairly indistinguishable at a distance, there are certain telltale signs when you get close enough, with size being the obvious one. I estimate he is about seventy centimetres in length, a figure I arrived at by measuring the space between two twigs on a fallen pine that he crossed near the weir. Halfway along he turned and looked straight in my direction. I had time to memorise his face, its broad head and long, narrow snout. He looked thinner. Haggard. Only those baleful eyes remained a roaring furnace of defiance.

Another day I find a blackbird limping on the ground with its eyes shut. It lacks the strength to take to higher branches and hops feebly away from me along the old railway into wiry caves of bramble, as if seeking refuge from the very air. Another scolds me and hurls past, diving beak-first into a moss-grown elder thicket. There are no melodies past the last line of gardens on Bilton Lane. Birds camp by feeders like refugees around cooking fires, hunched and hungry. Across all the fields and down the holloway I hear only one rusty *chip* of a solitary great tit. It is late afternoon under a high lead and gold sky and everywhere is bleak and empty, the temperature is the sort that robs your lungs of breath. Tundra air. Frost sparkles the spiders' webs between stems of dead cow parsley. Weather forecasters predict −10 tonight and the trees seem anxious. In the deepest part of the wood their trunks are starting to crawl with frost and they reach for each other with long, trembling branches. They know what's coming. Hardest hit are the insect-eating birds. The usual morsel-filled cracks and holes in bark are swollen hard with ice and yet it is a leaf- and seed-eater, a woodpigeon, I find dead on its back beneath a pylon. Ice has already softened the grey of its feathers into white and its face is a blur, like old fruit sagging with mould. I only notice it at all because of its two comically curled, pink feet, frozen stiff and sticking up in the air as though struck down in bed mid-prayer to the steel giant above. What I first mistake for a rook taking flight turns out to be a shredded black bin bag caught in the pylon's struts, tirelessly lifting and settling. I watch it for a while flapping in the wind until the cold becomes too much. In the backyard, cutting firewood for the stove, the axe tings uselessly off logs as though they are steel.

Later, once warmed up by toast and tea, I'm up a ladder painting a ceiling. The radio smacks about the bare walls, rebounding off windows still unsoftened by curtains. There is an interview with an art critic talking about how paintings supply the mind with an important 'fix'. Perhaps, but the edge-land provides a mental and physical transcendence greater than I've felt in any gallery. Merely the thought of it changes me.

I am drifting around the viaduct, frozen-breathed, following fox tracks. He must have been running: there are two prints, one in front of the other, then a gap and then two more. They lead down a gully to a scratched hole under a piece of corrugated steel, the sort my brother and I used to hollow out and commandeer in war games as kids. Fox holes. I lift the steel cautiously, then crouch inside. I'm aware that loneliness and the starkness of January are sending me ever inwards, into my mind and my memories. There is something about the stripping away of nature's decoration at this time of year that induces this kind of self-reflection. There is a trade, however; the earth exposes its inner-self too. Different perspectives are revealed each time I search for him. Buried things. New dimensions. One pre-dawn I sit and yawn and wait, close to the mouth of the holloway, down-wind, by an oak, under an intense tangle of branch, pylon and cable. It is the silent window between night and day, that slow shift in state. Liquid air. *Freezing.* The dark shrinks and disappears into the silhouettes forming in the west. My consciousness widens, rising with the night air, broadening with the dawn. Eastwards the sun, weak and rheumy as an old man's eye, hauls itself above the black trees firing the frost-fields into molten gold. The morning assumes a fragile blue hue, almost crackable, as transparent, triangular clouds freeze across the sky. Patterns appear on the surface too: the soft-focus haze of hedges blurring north and the corduroy shadows of tractor-combed earth. The edge-land is confessional, hiding nothing from me, revealing that which lies unwritten in books and libraries, unknown in the minds of those still asleep in bed. Those that have never seen this.

I watch the monotony of our constrained time unravel. The trees of the wood change colour with the rising light and trick my eye. Breaking out from the river gorge, their brown froth spills over the rolling curves of field and con-sumes the town. The spot where I am, the highest point in the fields, is suddenly an elevated mound in the heart of a royal hunting forest. Hoary old oaks shoot up shoulder-to-shoulder, sprouting dense canopies that turn the ground black with shade. Matted blackthorn, bramble and hazel unravel and twist in impenetrable lattices. Down through the woods, the Nidd licks at swooping branches of willow and alder. Heath and fern spring from its slopes; birch, holly, rowan and yellow-flowering gorse conceal fox, wolf, boar, grouse and deer. Huntsmen are here. They cut swathes through the virgin wood to the east using the edge-land's undulations and beck valleys filled with wild garlic to hide their human outline and cowl scent. I see all of this in a second and then, with a changing breeze, I am left with the glinting fields, pylons and the bare tunnel of the holloway again. The hunting tracks reform into Bilton's cul-de-sacs and estate roads – Meadowcroft, Tennyson Avenue, Knox Chase, Bilton Chase – the word 'chase' being the only indication of what came before. A young oak marooned in the centre of a farmer's field stands like a lost child after some natural disaster. When I get home a note in my book reads: *I love this place. It's the best place I've ever been.* Next to it is the scribbled drawing of the fox. I don't remember doing either of them.

I rattle around our Victorian terrace. I gloss cupboards, strip and coat walls and wonder at the histories I'm exposing and covering up. Where was the nail made that hammered in these floorboards? Whose were the hands that wore these ebony cupboard handles smooth? What did the mute, paint-splattered servants' bell by the bed sound like? Ghosts are filling the emptiness. These worlds of the dead and the living. Occasionally my mind drifts and I hear what sounds like the footfall of children running about upstairs or I'll turn and the half-light and my tired eyes conjure the shadow of a woman, hunched and carrying coal to the stove. I cook, eat dinner and call Rosie. *I wish you were here.* When she does return, fleetingly, she is exhausted from travelling and we are granted an all-too-brief window to fall back into our happy, human patterns.

As each weekend draws to a close we take to bed like one of us is leaving for war, pained by separation and seeking comfort. After she leaves I rise early and run to the crossing point before the world stirs. Passing the empty, spectral forms of buses on one dark morning, I read *2B: Bilton* glowing on their LED destination boards. 'To be' indeed. I'm surrendering to the edge-land, and it to me. My time is determined by the dusk and the dawn, by these fleeting moments of suspension between day and night when I feel most fully and wildly alive.

Another heatless, open-sky morning. I see no one, not even the buttoned-up mirages of dog walkers on the old railway in the early mist, driven from warm beds by the bowel habits of their pets. The meadow's grass is brittle, covered in a fine white dust and fresh with fox prints. They lead off in a trail so easy to follow it's as though there is something he wants me to see, something beyond the old railway and the mills, past the histories of huntsmen and the deer herds.

The last of the cathedral-deep glaciers melted here 11,000 years ago, but today ice sculpts the edge-land again. Trees are black lines scratched in blue and the air smells of wet, cold iron. White mountains of cloud are indistinct from the hill-line. Pylons twinkle top to bottom like vast river icicles. I hear the quiet waves of cars stirring on the main roads and see the moving chrome and glass shine in the distance like wet stone. There is a sacred calm and I imagine I am at this land's beginning, that very moment thousands of years ago when the great ice finally cracked and shrank back further north. I shut my eyes and let the sound of traffic morph into that of a flood river, breaking out from the glacier along a path of least resistance, tearing channels through the soft sandstone and carving out the river gorge ahead. The evidence is that humans reached here soon after the ground was released, conquering the rich, fertile earth as it burst with colonising seeds and spores after 100,000 years of incarceration. For so long there has been an unbroken line of eyes looking out across the gorge as I do now, seeing the skeleton tree canopies tinge with evening sun. The rift remains an open tear.

If you plumbed deep enough, you'd find foxes bound up in every layer of this land. Remains have been discovered in Warwickshire's seams of Wolstonian glacial sediments dating their presence here to between at least 135,000 and 330,000 years. Data reveals that when the great ice came again, some were driven south to more temperate climes – Iberia, Italy, southern France – others curled up in caves and dens and slowly turned to stone. But the margins were always in their blood. The species returned from exile as soon as the climate permitted, repossessing the fringes of the habitable. Their post-glacial remains have been found at several sites in Britain, evidence of a swift reclamation about the same time as humans. The crossing point for both of us was Doggerland, an earth bridge that once connected this country to Germany and continental Europe. Perhaps foxes were our guides then too and we followed them into this new realm. Certainly the flooding of Doggerland 3,500 years later isolated us both. We were trapped here together on the edge of the world.

It's been a few days since I saw him. The air doesn't help. It is a clinging curtain of cold, wet wool, clouding sight. Disjointed noises of machines and trucks rattle from the roads. My breath blows thick as a sea fret. I hear muffled shouts and, inexplicably, sheep. Closer is the clatter and chatter of jackdaws. All around is the feeling of confluence, of things happening just outside my vision. We're always told that time is linear, yet in this kind of atmosphere it feels more like a ball of string where points touch for the briefest moments and coexist in the same space. An overweight sheepdog sniffs at a gap in the undergrowth by the old railway siding before being pulled away. 'Leave it. It's dirty,' shouts a man, yanking the lead. 'Fox.'

I think nothing of ducking down and following the hole through the bushes. A squashed Fanta bottle lies by a torn clump of hen pheasant's feathers. An unusual dinner. Soft down is scattered everywhere, caught in cobwebs and brambles, but three or four beautiful, mottled, russet-brown wing feathers are still attached to a bony stump. Slightly curved inwards at the edges, it looks just like a little baseball glove.

The crossing point is shrouded in fog, which forms an impenetrable wall behind the houses, deadening distance. It creates the illusion that the world is only a single street deep, a wood-backed Hollywood stage-set in an American desert. But from the viaduct, a different vista. The view stretches twenty miles in one direction, revealing open country in an astonishing collage. The nearest fields are a British military green, then come the smudged lines of grey-purple trees and hedge. Furthest is a blurry ochre where division between land and sky can only be discerned by the crimson glow of evening. It blushes the low cloud as though a great fire rages over the horizon.

As the days pass I sense a thin, alarming energy rising within, like when you get a nosebleed and, head back, you swallow blood. All this time out here in the cold brings an extraordinary clarity; I have bursts of intense awareness where I can almost hear, see and feel too much. But still no sign of the fox. And no fresh kills either, or none that I can find. I wait nightly, though, camped down in the seams of undergrowth between the old railway and the meadow. As the hours fall away I feel no urge to return to the empty house. Instead, I sit and watch my hands changing shape in the falling light, my nails turning an iridescent black. There is a notion in the back of my head that if I can just stay here long enough, catch a glimpse of him again, follow him where he leads me, then I might attain a better understanding, a common consciousness with edge-land and animal. There is a noise and a movement across the meadow. I feel adrenalin flood my stomach as a breeze blows cold into my opened mind.

He senses me from the wood edge and freezes. I am shapeless, blurred by darkness and vegetation, but something. Something large. Something wrong. Disorder speaks: the way the misted tops of dead willowherb have been parted, a solidity among skeletal stems. *Hide*, it all says and he obeys. Dusk has passed swiftly, the last light flashing by. High-pressure sodium vapour flares through the town but only throws the surrounding land into deeper shadow, too dark to discern what lies on the far side of the meadow. He tastes the air. Scents swirl – *bark, pine sap, rotting leaves* – then a stronger taint, mine: chemicals and sugar. He knows it immediately; any wild fox would – *man*.

A hundred breaths later and we haven't moved. We are eye to eye, aligned under a sky flecked with stars. The ground between us is crisp with hoar frost, to my ears stilled but to his alive with the scratch of tiny claws. Wood mice trickle like rivulets through the under-grass. I sense the starvation that hangs about him like a cloak. The sound of scurrying stokes his hunger, but he remains concealed – hair raised, back rigid, body twitching. His heart beats with a fear passed from nose to nose for 3,000 years, greater even than the sour ache of his empty gut. Nose and black-tipped ears work to range smells and sounds. There is meat and garlic on my hands. To the east, a boar badger has blood on its snout as it defecates into a shallow hole, marking territory. A staccato fart from the town, the bucket exhaust of a souped-up Vauxhall Astra accelerating towards a red light, then slamming on its brakes. Somewhere a door opens, releasing the muffled beat of a stereo. At the same moment, the fox picks out the imperceptible brush of wing on branch as a tawny owl leaves its roost to fall on a shrew. These noises do not disturb him, though. It is the unfamiliar that breaks the deadlock. Across the meadow, cramp means a shift in my position, sending out the strange *swish-swish* of a waterproof brushing against itself. *Swish-swish*. Oddness, anathema to the fox. *Enough*. He slips backwards until his hind paws feel the incline of a steep gully behind. Then he turns and bounds down it.

A flash. And I go with him. We move, conjoined and flame-like, over the fetid leaves, dashing past an oak and into a holly thicket. A blackbird huddling on a low branch explodes in an upward flurry, chastising with a shrieking spray of notes. The leaping bite is instinctive but it catches only the waft of tail feathers. The fox grubs up a worm, chews it and waits for the wood to still again. After a moment, stealth re-forms like a membrane around us and we slip back into the trees.

Impossible, but I am following him still. An exchange, a fusion, has occurred. I suddenly see and understand. I know that it's seven years since his slippery birth under a gorse beside the river. I know that his mother and father were killed soon after his weaning, his father shot through the spine by a farmer; his mother's brilliance crushed by the glancing blow of a lorry taking sheep to slaughter. I see her laid out like a hearth dog among the silverweed at the road's edge, her tail wagging in the slipstream of passing vehicles. I watch him as a young fox foraging for beetles and shrews on fearful trips from the den. I see him grow in the summer that followed, becoming strong enough to fight off the foxes that came prospecting the edge-land as his father's musk faded from its fences and bushes. And I go with him now. I share his elemental possession of this ground, his mapping and claim via snout and gland.

From the Nidd to the old railway, the fox's nose bow-waves through field, hedge, meadow and wood. He knows all 320 acres by sight, scent and sound. He knows it when fat with mallard and basking in arbours of grass and rat's-tail plantain, the pineapple tang of mayweed astringent on the breeze. He knows what it is to forget the fear and doze among tufted couch strawed by heat, bees lumbering, pollen-laden, between the white funnels of bindweed as swifts sear through the blue above. But he knows it in winter too when all is hard earth, bleached air and burning bones.

Heels lifted, he paws through the wood to where it joins the holloway, rising south, uphill. Its hazel walls are blasted back by cold. Trunks look glazed with ice. Puddles pit the earth but their water has been robbed by cold, frozen into panes and smashed. He touches his nose to ghosts of plants, to cindered earth, bracken and bramble coated with rime. Only the hollies and goosefeet ivy have escaped this salt curing, their leafy pelts hanging glossy and green.

Halfway up he picks out a scent from an old run cutting over the fields. He takes it, heading west, over the plough

ruts, bobbing, sniffing, detecting. A roe deer print is gouged
into soil beside a rusted door hinge. He investigates three,
four, five more, the last splayed where the deer broke into
a run and its hoof took the weight of muscle. Further along
is the dark stain of frozen blood. *Rabbit*. He gobbles a
severed foot and the skin and head of a young buck killed
the night before by a badger. Then he scratches around a
slab of stone, a fallen gatepost for a path long forgotten,
scrounging beneath for chrysalides and seeds.

At the edge of a field the ground swells to greet a boundary hedge. Bare hawthorn and blackthorn comb his coat as
he twists beneath, sweeping for fruit, but mice have raided
the last of the larder; even the frostbitten clusters of rose
hips and haws are gone. The thinnest twig tips tremble and
squeak against each other – *cheep-cheep* – anticipating the
calls of warbler chicks that will explode from these hedges
in spring. The fox rests in a clump of hogweed, unaware
that it was once a Neolithic knapping point; two metres
below, Kentish flints pepper the ground. Above, stars spin
around a new moon. Breath freezing on his snout, he blinks,
sniffs and scans fields awash with pearlescent glow to the
west. Hemmed by dark hedge and wood, they fold into one
another before succumbing to the sprawl of the town, a
black sea flickering with phosphorescence. He sees a million
eyes: street lamps and headlights, the yellow, maggot bodies
of commuter carriages screeching, hissing and rumbling
back from Leeds and York. It is an ever-respiring beast that
puzzles him. He fears it; he craves it too.

Lean, hunched, he roams along runs that resemble the
eroded ditches of dry rivers. All these tributaries loop eventually back to the meadow edge where, cautiously, he sniffs
for me. He trots over the icy tufts, springing a bank vole
from stillness and capturing it by the legs. There is the jerky

snap and click of sharp, yellow teeth through bone. Then he drops it, puts his paw on its head and tears it in half. Somewhere deep in town an ambulance flicks on its siren.

Haow. Haaaooow.

A different call. Animal. *Close.* It's warm and wide-throated and, head raised, vole wobbling in his jaws, the fox feels it more than hears it. *Vixen.* His ears twitch and range. She's young and in his territory, down by the viaduct. Over the old railway the fences of the housing estate reverberate with a volley of barks. A single German shepherd triggers the half-forgotten instinct of the wolf pack, sending a ripple of snarling and barking through the houses. Claws scrabble at kitchen doors and garden gates. The vixen ignores them and sings again – *haaaaoooow.* The sound pierces double-glazed windows, stopping forks halfway to mouths, wondering at the scream outside. It fires the urge to mate between the fox's thin hips. If he'd been stronger, perhaps he would have sought her out, but not now. *Haaaaaooooooow.* Others will be coming soon, young dog-foxes with only three summers under their pelts, slipping out from fields and the wastelands behind the playing fields, warehouses and paint factories to come here and search for her. The fox knows to encounter them as he is, flower-frail with hunger, would be dangerous. His patch could be taken. Or worse.

Territory is everything in winter. To be forced to roam in the open would be fatal. The fox hobbles off and sprays the brambles around his gully. All must be marked. From his position on its ridge the trees below seem to collapse inwards, caving in on the winding watercourse running along the little valley's bottom. Bilton Beck rises here and burbles over a silt bed strewn with black stones all the way to the Nidd. In places plastic bottles spin endlessly in the eddies; in others, ice sheets join the banks, roofing the stream. It rarely freezes completely, though, and he knows this. No wild fox would dig a den where there wasn't a pond, marsh or stream within a few hundred yards. Not that he dug this one; it is the remains of a badger sett hollowed beneath a beech a hundred years ago. Floods and landslips have exposed the tree's roots, leaving them clutching at the soil like a sparrowhawk's talons. Behind there's a scrape where the fox lies on warm days surveying the gully, but with the air sharp as a thorn in the nose, he disappears into deep earth, ducking along a root-draped passageway towards the furthest chamber where the ground is soft, black and nitrogenous. His copper fur parts to reveal a pure white undercoat as he flexes his body into a curl.

The soil is a repository of old smells and they come to him, given life by warmth and movement. The strongest is his vixen. For a moment he remembers licking, the wilful submission of mating and the sweet tang of kits. The den has known many such balls of brown fur squirming in its earth for heat and milk. In his drowsiness, time past and present combine and soft, clawless paws clamber over his face. Blind liquid eyes push up to his. His fur stirs with the hot, sweet breath of pink, mewing mouths. Then he is alone again. He dreams of root, burrow, earth and blood.

The fox's mate abandoned the den two years before, just as tender goose grass and nettles began to carpet the under-brambles. He watched her trot to the horizon at the top of his gully. Pausing on the ridge she became a silhouette under stretching arms of an ash then disappeared over the edge. The kits bounded after, scrabbling up the slope in pursuit, white-tipped tails flicking, fur only just fawned, all barely thirty days old.

The foxes had mated together before, raising litters that grew fat, first from her milk, then from the rabbits, wood mice and pigeons he'd hunted. But hunger came often in that last winter together. Sixty years earlier, farmers had thrown rabbit corpses infected with myxomatosis into the burrows that edged the wheat fields to the north. The infection spread ruthlessly, decimating warrens that had been tunnelled under the wood for a thousand years. It was a grim plague that still haunted the survivors' descendants, flaring into epidemic proportions during hot summers when the breeding conditions were perfect for the virus-carrying rabbit flea. That year it had thrived, spreading from coat to coat so that by the time the beech over the foxes' den had shed its last leaves, the prey they relied on was in sharp decline. The fox hadn't even needed to hunt the few rabbits he came across swollen, shivering and foul with disease. Eyes bulged red and bloody from skulls, sightless, scratched out. Blind, dumb and disoriented, they dragged their useless hind legs in pathetic crawls for cover; he felt none of the joy of execution when stooping to break their necks. Even the meat tasted poor, the flesh corroded.

The foxes hungered, a pain compounded each night by the smell of food drifting over the meadow. Sickly thin, hunting in the frozen, misty margins, the fox bit at the scents – *chicken bones, meat, hot marrowfat, rotting vegetables, baked wheat* – and scrabbled violently at the earth in his search for worms and insects. Once he dug up a squirrel's cache of acorns and ate them all, carrying none to his mate. But despite the gnawing in his belly, he kept to his territory, never straying beyond the old railway. Being a wild fox, the smell of man triggered received fears, memories that had passed from fox to fox via trembling whisker and womb. His were blurry visions of The

Bramham Moor Hunt, founded in 1740 by the improbably named MP for York, George Fox-Lane. Although long since merged and moved to more respectable pastures, it had once been a regular sight through these fields and woods. The fox half-remembered things he'd never even seen: spectacular horses, duns and greys, gleaming horns and gentry in blue velvet jackets thundering along the treeline, driving piebald foxhounds up the gully's sides; thick winter fox pelts skinned from pink carcasses left to rot in snowy fields. His vixen had been different. Littered in an old construction pipe behind the sewage works at Bachelor Gardens, she'd scavenged discarded takeaway polystyrene and bin bags since weaning. She'd learned to wait from the cover of parked cars until closing time brought the rush for takeaways that would be spilled onto pavements by drunken hands. The smells of man compelled something different in her: *Leave. Feed. Mate again.*

The fox had waited for her return, the freshness of rubbings around the den keeping her alive in his snout. Each night he patrolled, marking trees, leaving his twisted black coils on stumps and surveying the meadow from below the same ash tree his ancestors had. A favoured spot, its bark had been worn smooth by generations of foxes drawn to rub there by a usefully positioned nail. This was an unintentional memorial, hammered into the trunk in 1914 by Lieutenant Thomas Watson before leaving for Egypt with the Leeds Pals, with a vow to his fiancée, Elizabeth, that they'd remove it together when the war was over. That benediction never came. The fusion of bone and mud they salvaged from a shell hole three years later was spaded into the earth at Tyne Cot Cemetery, Belgium. The nail was left to the tree and over the years became almost consumed by bark. It stuck out just enough to snatch a few hairs whenever the fox scratched against it, forming a tuft that was foraged every spring by blue tits to line their nests.

Moons passed, five in all, but with no sign of his mate. As dusk heralded the sixth, the fox prowled along the river past the collapsed bank of his birth den and surprised a young rat, devouring it in seconds on the crescent of a small muddy beach. He cracked its skull with his back teeth and swallowed the tail. Strength surged through his limbs and he lolloped up the gully side following her trail as a flare of sun sharpened the horizon into a clear line. The day had been numb and grey and he trotted towards that fading frequency of warmth, weaving, nose to ground, tracing her scent through the fallen branches and infant snowdrops. The meadow was growing then and among a swathe of grass and sprouting dandelions, a cock-pheasant poked up. The fox sank to its haunches but the wind changed; a breath lifted his hind fur and the pheasant rattled off in a volley of clucking that echoed through the wood.

The fox crept along the treeline until he reached the old railway. His mate had paused there to wait for her kits, in the same spot where a signal box had once stood. He smelled them on the corner of an old brick poking through the mud. At this distance the town looked different, like an open mouth. Houses loomed. Behind them, the blurry amber curve of street lights. Power lines crackled off south-west towards a groaning electricity substation. Metal smells. New patterns and shapes. Drizzle began and he knew it was the dark precursor of a storm. He smelled man. Then, with a step, he left his territory for the first time.

Breaching a privet hedge, he paused halfway through the orange wash of a rainy cul-de-sac. Its reeks disoriented: oil, tobacco and mint smeared into pavements. Rain drummed on street-lamp casings, swirling down from a sky the hue of blackberries. He felt the tarmac tremble and turned to see moving lights closing at speed. A screech and a long horn-blast sent him scampering along a ginnel between two houses. Leaping over a wall, he slipped down a bank and into a scrappy wood. The storm was growing, awakening the earth; wind stirred the trees. The foot of the slope blew with plastic bags and bottles and a line of elder bushes entwined a twelve-foot security fence. On the other side lay a railway, a dull bronze line that thrummed under a row of arc lamps. The hissing whip of a train ran along its length, metal against metal, an unfathomable weight approaching. The fox slunk away, trotting southwards, his black paws splashing through run-off. It was then he saw it and his body jerked and froze: fur flashing. Red-brown fur. And that smell. *His mate.*

The vixen pushed through the wispy curtain of old man's beard across the railway on the opposite bank and, lit up by the corollas of light, slipped onto the tracks to chase a crisp packet. The fox's ears flicked at the noise growing louder, a sawing sound like a gull's shriek that shrunk him down into a crouch, yet his vixen stood her ground, watching something with a foreleg raised. Hunting. Then she pounced on a rat as it fled a flooded hole. Suddenly that moving wall of sound and weight and light was upon them, over them, part of them. A terrible flare and scream.

The fox ran, his legs a mess of speed, a force pushing him onwards along the fence and into the edge-land. He ran down that wild corridor and over the crossing point until his strength gave way and he curled up beneath the last elm on the river to sleep. By the next evening his heart had slowed.

*

He wakes now, disturbed by a vole scampering to its tunnel in the den wall, and stretches stiffly in the dark, legs numb from being too long motionless. With a yawn he tastes disordered air. Through the rooty aperture of the entrance, silent, feathery flakes swirl this way and that, aimlessly, diagonally. He blinks again, nibbles his fur and sniffs. *Nothing*. All scents are buried now. Even the voices of birds are quietened by blizzard. Pain swells in his arthritic legs. He has never been hungrier but he knows that hunting would be futile in falling snow. Covering his snout with his brush he lies awake, waiting, weakening.

By dusk the next evening the world has become monochrome. Lines of rooks huddle in the fields searching for a gap in the white crust. The fox is at the holloway, stalking a crow as it smashes its beak into a chalk-hard drift. With every strike it sinks a little until it raps impenetrable earth and draws out a rattling, *krah-ah-aahhh*. The fox slinks on his belly, but the crow hears and hops, skips and vanishes into grey air. High above, the drone of a great metal bird whines and throbs, retracting its landing gear to begin its migration south.

Despite the snow, moisture in the air carries sound and scent: a foggy hum of lorries and cars and smells pumped from industrial vents. A Chinese takeaway on Bilton Lane is deep-frying pork. The fox's ears twitch. He snaps at the air, retches, coughs and yawns. Disturbed, the rooks rise from the fields and throng the wood. Woodpigeons flying in pairs fold back their wings and come in to roost. He knows birds are beyond him now and he resorts to his clump of dead hogweed at the field's edge as the street lights of town begin to shimmer in the darkening mist beyond. After twenty breaths, he rises and limps towards them.

Behind the old railway, down the siding bank, his territory finishes by a clutch of garden-escaped hellebores. They are a purple marker among the ermine snow. A rotten garden fence blown outwards lies semi-buried by earth piled into mounds from rabbit excavations. Moving from hole to hole he thrusts his head in and tastes the faint air of a warren. Memories flash: fur against his tongue and teeth, flesh, the breaking of bone. The scents are faded though and the warren collapsed. A Staffordshire bull terrier killed the last breeding doe a year earlier as its owner stopped to light a cigarette. The clutch of pink infants left orphaned in the burrow was a feast for a pregnant rat. The fox smells it all and moves on.

Crossing the collapsed fence, he crawls up a bank and under a laurel into a long yard thick with snow. The crystalline layer has turned plant pots into giant puffballs. A wheelie bin is on its side, lid open and overflowing with rubbish; the stench fills his nose. It is this that has drawn him. But there is another – *dog*. It's weak, though. There's been nothing in here for days save for a pair of blackbirds investigating another long-empty feeder. Only their spidery scrawls flaw the snow's luminescence. The fox creeps, body tensed, each boot barely shifting the surface. Pad, pad, pad, he creaks towards the square of the house looming, cliff-like, at the end. A blind in an upstairs bedroom glows yellow. That's all. He pauses. *No.* There is another, an intermittent flashing like sunlight on the river. It is coming from around the corner of the wall, near the bins. Suddenly the house wall erupts with a gurgle of steaming water swirling down a drain. He smells something familiar – *elderflowers*. He hurries into a trot and rounds the house, bringing a bright conservatory into view, a television blaring at its centre. In a heartbeat he is down on his chest, ears back, baring his teeth. Every sense tells him to flee. Here is all he despises: uncertain ground and dangers, but his hunger is like an animal eating him from the inside. He darts for the wheelie bin and tears opens a plastic bag like a rabbit's stomach spilling wet cardboard, eggshells and half-eaten vegetables. Burying his snout deep inside, his teeth clamp on a chicken carcass, which he drags onto the snow to crack the cartilage and swallow the pale, forgotten under-meat.

Desperation has lowered his guard and he doesn't notice the young dog stirred from bed and walking to the conservatory window. Outraged by his wild form, it slams its paws against the glass, barking sharp violent rasps. The sudden scream from upstairs sounds like a rabbit caught by the neck and the house explodes with yellow. The fox bolts the way he came, running hard, oblivious to the burning pain in his bones and thinking of nothing but reaching his territory, his earth. Yet he doesn't run in its direction. His instinct is to lose enemies on foreign soil and so he doubles back, turning west along the old railway, dashing through borders, over fences, up bramble banks and through scrappy undergrowth strewn with fly-tipped glass and metal. Under a hedge he smells another dog-fox – *young, strong* – but presses on, past an allotment's drunken fencing, flushing woodpigeon as he runs. A hundred yards further, behind the slide and swings of a children's playground, a snowy belt of woodland leads down to the river. Scenting open water, he slows to a limp and skirts a shallow pool frozen and flecked with litter, ringed with the prints of moorhens.

Watching him unseen from the brow of a bank is a two-year-old dog-fox who tracks the bony shape panting beside his water source before emitting a high-pitched whine of warning. Surprised, the old fox jumps, spins round and retches back a gekkering call. His eyes are wide, muscles tensed, but the younger fox has no urge to fight him; he knows the intruder is too weak to contest ground. He stinks of fear and of death. But he carries another smell too, something familiar – dens and warm earth. The young fox lets out an ululating noise and lets his father pass, following his hobbling shape until he becomes invisible in the trees.

Moon renders the Nidd's surface polished silver. The fox breaks it with his tongue and laps. There is no other sound than the continuous, indefinable whistle-hum of the sewage works, a noise he has never heard so close before. Instinctively his senses orient him and he turns east towards his territory. Mist haunts the river edges, skirting the alder and snow-dusted pines. Nosing his way downstream, the fox creeps through the cover of dead vegetation until his own faint scent drifts over a low woody rise. As he crests a trackless path the night-veiled land beyond begins to resolve into the recognisable shapes of viaduct, wood and meadow. A sniff. *Close now.* He quickens his pace, limping through a thicket of hawthorn and up to a leaning fence covered by an impenetrable wall of creeping bramble. Rather than retreat, he darts through the narrow gap between a fence post and tree, landing in a ditch on his front paws. Immediately he knows he's been bitten.

There is a scratch and the feeling of being gripped. Sharp teeth hold his back legs. He yelps and coils round, snapping and hissing, fearing badger or dog, but his jaws clamp on something worse. Old discarded wire, barbed and twisted into loops, is buried deep into the flesh of his left hind paw and noosed around the other, suspending both in the air. He pulls and chews at the wire, dragging himself across the ditch, yet only deepens the wound. He tries jumping back and forth, then scrabbling over the fence from the other way, but the noose tightens and twists with the efforts, soaking the snow with blood. Eventually, exhausted, he collapses onto his side, ribs heaving, too weak to move.

The little ditch is massed with night when a robin breaks the silence – *twiddle-oo, twiddle-eedee, twiddle-ee* – a downpour of notes that wakes the fox. He blinks up at the song as it ripples through the wood. The robin is anointing the earth from a branch of a nearby crab apple, legs braced, fiery breast puffed and facing the red edge of the rising sun. Another sings, faintly, from the housing estate across the meadow. The edge-land is stirring. Slowly, trunks and branches become etched in a cream sky; light blooms in the east. Starlings settle again on the pylon cables. Rooks pace the fields. The *wish-wash* of cars rises in a thin line.

There is no pain, only numbness, and for a moment he forgets his bonds and tries to stand. His yelp frightens away the robin. Licking, pulling and biting the wire again agitates the cut further, for the old post it is nailed to is crooked but strong, a creosoted chestnut pole driven deep into the ground by a farmer who believed a job worth doing was worth doing well. More fretting flares the wound; it boils up with fresh blood until it's too tender even for his tongue to touch.

A frozen puddle a yard away half-melts in the midday sun, maddening his thirst, but all through the changing light of the mauve and silver afternoon he can only scratch at the ground and bite at the snow-topped brambles. Soon every breath is a wheeze. As the sky bruises, he hears the continuous exhale of cars returning. The puddle thickens and solidifies again. A rabbit kit appears at the top of the ditch. Curious, it approaches the fox and becomes confused by the absurdity of its trussed legs. The fox blinks; the rabbit has changed into a rook, assessing him, preening under a wing, cocking its head and letting out a loud *kro-aa-ak*. Even hisses and snarls don't scare it; then the fox finds he can neither uncurl his lip nor lift his head.

Fringing the ditch are twelve snow-rooted silver birches. They look like apparitions, hard edged but soft-skinned, the luminosity of winter landscape distilled in their cream-russet trunks. The fox thinks briefly of his den; then forgets everything. The rook hops about the lip of the scrape, bending closer, twitching, and ducking back in feathery leaps. Others prowl too, a silky crow and a pair of grey-hooded jackdaws. Through a tangled web of purple birch twigs the firmament changes as the sun becomes a rim on the south-western skyline. Clouds gilded by the dying orb hesitate for a moment in a forget-me-not-blue sky, before it changes to the hue of ripe wheat. Amber comes next, before daffodil and rose, then everything assumes the dark crimson of field campion. This stretches the furthest, reflecting in the ice and snow, turning them bloody. Fading, it leaves only the husks of trees and hills, houses and farms and the smouldering black stems of wind turbines on the horizon. The last colour is the shrinking topaz of the fox's eyes as his breath dissolves into the darkness.

A month passes, maybe longer, before I find his body. It is a bright, fiery morning, cold and sharp, with snow still on the ground and trees as I wander west from the viaduct, down towards the Bachelor Gardens Sewage Works. I have half a mind to photograph the river, which is shining like tin foil under a climbing sun. Near a cluster of silver birch I pick my way around a ditch overgrown with bramble. Clambering over the last vestiges of a fence, I smell it – a strong, sweet, rotten smell – and glance down to my right. It takes a moment to make sense of the mess of body lying at the bottom of the scratched trench. The fox is sagged, sodden, blackened, caved-in, his back legs stripped to bones and still trussed up in wire. It's heartbreaking to see and made worse by guilt: I'd almost forgotten him in these intervening weeks. I searched for him after he disappeared, looking in different places and at different times, but in truth I knew his work was done. I am beyond the crossing point now; the edge-land is open to me in a way I couldn't have understood before. Following him has deepened the map, unfolding it, throwing it into relief. Stories are within reach. But the human world has been pulling at me too. Rosie has moved up from London. Work is picking up. Other exciting news has thawed my loneliness. And now, looking at the fox's sad, skeletal shape below, I don't see psychopomp or guide, but a wild animal again. His face is a shrunken and squashed mask, all life pecked out of it; his claws are broken from trying to tear himself free. That flame fur has long since leached into the earth.

When I was twenty-four, I was rushed into hospital for an emergency operation. Three hours later I woke, still blurry with anaesthetic, to find the surgeon who saved my life standing at the foot of the bed, holding the appendix in a clear tube. 'You're a new man now,' he said. 'I'm giving you back to the world.' It was a strange moment and a strange turn of phrase, which is probably why it has stayed with me. Seeing the fox elicits a similar sensation, like a part of me removed. At once an ending and a beginning.

Sun rakes the edge-land. New light over this now-familiar ground. It feels so long since the world provided any heat that I put away my notebook and lie down, using my pack as a pillow. Beside me is a patch of coltsfoot, an inadvertent little graveside bouquet. Its flowers, which bloom before the leaves, are the nourishing yellow of free-range egg yolks – an assertion of life. And all around, the sun bleaches buds of birch and hawthorn, its elevation such that it seems to spotlight everything: the edges of stems and stalks, a passing pigeon's wings, the rooks *yak-yakking* in the meadow, a lone bumblebee. The curtains have been ripped opened and the dustsheets dragged off. Lying here with the warmth full on my face and the *drip-drip* of ice and snow thawing in quickening rhythms, I can almost feel the earth turning.

ULTRASOUND

There are owls everywhere I look. Cut from felt and sewn onto cushions, plasticised and stuck on car windows, printed on clothes, curtains and lampshades, encased in children's keyrings. Walk along any high street and they line up in the furniture and fashion stores, turning retailers into cartoon bird sanctuaries where hawk, eagle, tawny, barn, snowy and great grey owls mass behind the glass. There's even one in the hospital. Between two blue fabric boards pinned with notices and leaflets, the local primary school has decorated a section of the Antenatal Clinic's wall with cut-out animals. It's an odd menagerie: tigers, elephants, lions, a giraffe, a fox, a red squirrel, a hedgehog, and then what looks to be a tawny owl. I say 'looks', because it is not much more than a splodge of brown and cream poster paint, but two wide, front-facing eyes give it away. Little hands have captured a subtlety that fashion designers and toy makers usually miss: an owl's face is curiously human.

After tingling our hands with the gel dispenser at the door, Rosie gives her name to a lady at the desk. 'Your twelve-week scan, is it, love?' she asks. 'Take a seat and someone will be with you soon.'

We sit on a line of smart, red chairs opposite the solitary owl as it stares out from the branches of its broccoli tree. Our fingers are entwined under the handbag in Rosie's lap, wrists touching, pulses dancing in nervy excitement. She sips water occasionally, to try to quell the ebb and flow of morning sickness. For good reason the Antenatal Clinic is a long stretch from the tobacco fog of smokers at the entrance and the baked-goods and coffee smells of the hospital café. You have to follow a red line down numerous shimmering square corridors. Once inside, this ward seems softer than the rest of the hospital, lighter and more relaxed – all peaches and pine. This is the good end of the human journey, I suppose; a place where the body conforms to its predetermined trimesters and where pain has a purpose at least. And there are plenty of reminders of the prize if all goes well: posters to promote breastfeeding show happy mothers cradling cute babies; cherubic faces crawl across white pillows in adverts for local photography studios.

Had the clinic's windows faced the outside world rather than an enclosed courtyard, we might have seen the bright, cold morning and its sky of plate-glass blue. Instead, the waiting area's luminosity comes from wire-waffled squares of fluorescence in a tiled ceiling. Wooden toys stand on low tables by racks of tatty magazines that I'm apprehensive about touching. So we sit and look at the owl, stroking each other's wrists with our thumbs and listening to the muffled beeps, buzzes and squeaks of shoes striding across disinfected vinyl.

The sonographer, Rachel, is pretty with hair tied back and wearing a fitted white tunic with blue bands about her arms. She carries a clipboard and a warm smile, but is clearly up against it. 'Would you like to follow me, please?'

We do, to a small but bright room, just enough space for the equipment, a bed, sink, storage and the obligatory multicoloured bins with toxic warnings. It smells clean, surgical, like the hand gel at the door. Rosie lies on the freshly papered trolley bed and rolls up her top, as directed. Rachel turns a pole on the room's Venetian blind, plunging us into dark, and seats herself at the ultrasound's kidney-shaped plastic control board. With its big buttons, detachable bits and ergonomic curves it looks strangely childish, a Fisher-Price 'My First Ultrasound' toy. That is until it fires up and the screens come alive. Serious beeps and lights. Rachel takes control like a pilot at a console. She looks back at me as I stand nervously, holding our jackets and Rosie's bag.

'You might want to sit down.'

Yes. Take your seat. Relax. Prepare for the in-flight movie.

I hold Rosie's hand as Rachel squirts a big circle of thick, bluish gel over her stomach. 'Sorry it's cold,' she says as Rosie hoots. But it is excitement, not discomfort. From where she's lying, she can't see the monitors: Rachel has both of them turned towards her. I imagine this is in case of problems with the foetus; they spare the parents the emotional connection that would be forged by the act of laying eyes upon a life. But I'm tall and nosy so I lean back my chair and get the first views as the torch-like transducer probe nuzzles its way around my wife's belly, ranging, listening.

Black and silver storm clouds tumble past, swelling and contracting like squeezed balloons. It takes a second for Rachel to locate the uterus, a throbbing mercurial ring, and then hover over the placenta. She operates the probe expertly, changing direction, clicking its sides and zooming in for a better view, closing on the blackness of the amniotic sac. What I see is even more black and silver circles extending, opening, parting and joining, making shapes. There are all sorts of landscapes forming from Rosie's workings; we're flying over the Lake District's craggy northern fells, winter moors, domed Salisbury hills and above the moonlit Thames – all the places we've been together. She carries them inside her. Rosie is watching my face; I squeeze her hand and smile. Then, when I look back at the screen, the snowy pixels have arranged themselves again, forming, for a second, the distinct face of a tawny owl.

'Well?' Rosie whispers. 'What can you see?'

You don't usually find tawny owls (*Strix aluco*) in hospitals, although their habitats are changing. They like to roost in woods and the high crevices or branch cavities found in old, deciduous trees, like oak. In our endless skirmishes over land, woods continue to disappear and the felling of dead, hollowed trunks has seen a denuding of traditional nest sites. But being a year-round resident, tawnies hate to concede ground and the stability of their numbers in the UK is down to a willingness to take alternative accommodation in order to hold territory: purpose-built owl boxes; squirrel dreys; unattended crow, magpie or heron nests; even dilapidated buildings. As a student in Leeds I once heard a male calling from a windowless warehouse near a motorway on the long six-mile walk home from a party. Mostly they seem to love the abandoned places, the unmanaged islands where man has temporarily laid off interfering and allowed functional ecosystems to thrive – forests, cemeteries, churchyards and, almost always, edge-lands.

I've been watching a pair of owls down in the wood for weeks. About four weeks, to be precise. And it's probably truer to say listening for rather than watching, as being nocturnal and soundless in flight, they've been almost impossible to see. No matter. Their calls sneaked inside me, lingering in my ears as doggedly as fag smoke used to hang about your clothes after a night in a pub. I first began hearing their duets in the wood when I was out looking for the fox. At times they were frightening, like the panicked, gurgled screams of shipwrecked sailors drifting in a black ocean; at others they were the cooing comfort-sounds made by a new mother. Each call carried the natural reverb of the river gorge, lending them a peculiar 'Wall of Sound' sonic resonance, like the harmony part on 'Be My Baby'. Roomy. Spacey. And those shrill, terrifying shrieks and low, loving hoots kept me company at night, growing ever louder in my re-ordering world.

Horizons began to widen in early February. They always do. It is something about the lifting light and sky. Days no longer seem so abbreviated, in such a mad rush to reach their conclusion. Even in the architectural confines of central Harrogate one afternoon, I saw hundreds of high, hollow rectangular clouds stretching off indefinitely with the bumpy texture of old oak bark, filling the air with a psychedelic mauve. *Look up*, I wanted to say, but no one seemed to notice. People moved from shops and offices to cars bearing mobile phones or stood bored and smoking, playing Candy Crush as they waited for buses. The evening suggested limitless potential, so I walked home past a row of once-pollarded ash growing wild behind a high garden wall. Birds were on my mind already. Robins, wrens and blue tits were out of hiding and contesting territory with such fierce beauty that their calls drowned out the passing cars. Held in their brief tractor-beam of song, I tried to follow each bird's movements as, every few seconds, one flashed up to the wooden mouldings of a Victorian gable, then dived back into the fray. The sky bled into a soft coral at the crossing point and the black, broomstick-tips of trees along the old railway fanned like capillaries of the heart against it. It was an unearthly window, that changing of the guard – cold, clear day to bitter, black night.

By half past five the light was little more than a blaze of red through the cruciform trees. Dark took the gorge unchallenged. Fallen branches were brittle as antlers but little repositories of life spotted every living shoot of tree. Rock-hard, egg-shaped and bright green, each was an assurance of spring, the promise of life lying dormant. Buds dotted a nearby oak branch. These were different: rounded, rust-coloured and massed in clumps at the end of each meandering twig. A tall beech had pointed copper arrowheads growing from its elegant curved boughs. Perhaps it was just the illusion of their fawn hue, but the leaning pines by the river seemed to retain the faint heat of the day so I hung about their trunks as the last glass bottle-blown notes of a woodpigeon faltered and ceased. Then the river glimmered with the echo of the owls. That extraordinary, aching call. Except it's not 'call', singular. The famous *tu-whit, tu-whoo* of children's stories is actually two birds communicating. The female utters less a *tu-whit* and more a *ke-wick*, but even this seems a poor transposing of her brief, piercing cry. Similarly, the male's hoot is not so much a *tu-whoo*, it is more a syncopated *hu ... hu-hulo-hooooo*. Warmer than it reads. What the simplified reductions miss are the dexterous parts, the deft little trembling descent at the end like a folk balladeer adding emotional gravitas.

Such vocal attention to detail is hardly surprising. Sound is everything to tawny owls. They exist in loose communities but there are strict rules about spacing and tenancy. Territories lock together as neatly as housing plots delineated by the Land Registry, but having no sense of smell (or Land Registry), theirs is a predominately aural world. Calls are territorial assertions – *this is my hunting ground; this is my mate* – and like a catchy chorus, they are infectious. Hooting leads to hooting. In the pauses between them, I could hear the faint echoes of other pairs coming from downstream and from the west towards the meadow and town. I imagined the vast linguistic conversations that must flow out, around and across our night-drugged world, the silvery webs of chatter via which disputes are settled and breeding determined.

Books tell us that precisely in the same way we can recognise a change in tone in the voices of friends or family on the phone, owls living in proximity know one another through minute differences in vocalisations. Should a male fail to respond or its call sound weak, word will get around and its territory quickly snatched in a coup of chasing and hooting. But their singing is erotic too and, in established pairs, a male's crooning stimulates ovulation. The two I could hear had almost certainly paired in the autumn and were probably well into the feathery tangles of mating. Or perhaps even past that, relaxing through post-coital rituals of preening, rest and roost, pressed up flank to flank by their nest site. There's a softening in behaviour after copulation, a shift away from the talon-flashing late-night flirtations towards mutual trust and friendliness. Maybe the calls I could hear weren't dirty talk; they'd moved on to rowing about kids, mortgages and money.

Although I didn't see or hear it move, the male's calls were suddenly directly above me in the black crown of a pine. I'd never heard an owl as clearly or closely, at the level of valves and throat vibrato, of air being worked to an owl's purpose, but I couldn't stay much longer. My own mate was calling; Rosie had been at home in bed for two days with sickness and exhaustion. Now she wanted feeding. The text request was unequivocal: 'please bring pizza'. It didn't sound much like the flu to me.

A coincidence: snow fell the exact moment I learned I was to be a father. And in that moment everything changed. Entranced by a blue line emerging in the little white window of a pregnancy test, we looked up to find snowflakes pouring into the street as if tipped from the back of a truck, obscuring where the horizon meets the grey roof slates of the terraces opposite. Then we lay together on our bed and watched the snowstorm form. Squalls of silent white swarmed the glass then backed away; flakes doubled in quantity, thickening into conjoined masses like cells dividing. The town was soundless; traffic frightened and slowed. *Remember all of this*, I told myself. *Remember this ethereal quiet and the lines of melting snow streaking down the glass. Remember the heat of Rosie's happy tears dampening my shirt.*

After a while, a burst of laughter from guests downstairs popped our bubble. Suddenly our house was filled with life. Friends from London had booked to come up weeks earlier before any of this was on the cards. It was too early to tell them so we pretended our giddiness was down to the snow and suggested a walk in the blizzard. Half an hour later we were wrapped up and battling down Bilton Lane past

houses that looked offended by the covering, like pension-
ers on their way to church unexpectedly caught up in an
impromptu foam party. Cars were crippled. The wind
rushed low and fast from the north and sent waves of snow
upwards so that the storm appeared to be emanating from
the edge-land. A sky of white tracer and the land beyond
thick and grey as putty. It soon reached its zenith, though,
and waned as we reached the crossing point. The familiar
topography beyond formed slowly, as if through a de-
misting shower screen, then the low clouds flared with sun
and evaporated into sky. Suddenly, in every direction, there
were two tones: the linen-white of fresh fall and the coal-
black of under-tree and under-hedge. Chaffinches struck
up from hazels as we squeaked up the lane through virgin
drifts. From the high rise of the fields the sun spreading over
the distant reaches of the landscape made my eyes ache, but
I fancied I could see further than ever, an effect wrought by
this great levelling. All bumpy ground was smoothed; there
was a new coat of paint. Everything clean and clear.

We let the others go ahead and stopped at a gap in the
hawthorns at the holloway's entrance. The edge-land was
still new to Rosie and the view had rooted her to the spot.
I put my arms around her, over her stomach, and my head
on her shoulder so our faces were side by side looking at the
same things – the lone oak, the razor-cut line of hills, the
pylons, the smothered steeples, domes and towers of town.
My mind raced with the power of nature inside and out;
joy swelling and fluttering my stomach, happy as a man can
naturally be. But it came with worry. I couldn't shake the
thought of how easily things can go wrong. 'Wrong' is not
the right word, right and wrong being human concepts;
what I mean is the sense of how things sometimes turn out
differently from what we hope. Nature is impervious to

wishes. Cells fail. Life wanes just as suddenly as a snow-storm. I said none of this, of course, but stood there taking everything in. What I remember most is the sound. The sheer, beautiful absence of it.

Turns out that sound is a pretty important part of the twelve-week scan. They don't tell you that; the emphasis is always on what you see rather than hear, but the probe is a microphone too. Rachel leans forward, flips around the larger monitor and turns up the volume. The speakers cut in with a hollow *cuurrrrrr* – the static of a detuned radio or far-off industry, like the idling sewage works heard from the viaduct. Rosie's grip on my hand tightens as the probe sits tail up in the air, nudging into her stomach. The sound becomes louder and punctuated by a quick, pounding, sluicing rhythm. A heartbeat: *woosh-woosh-woosh-woosh*. The whole thing is weirdly machine-like; I think of Second World War films when you can hear a ship's propeller approaching from inside a submarine. On screen, the grainy silver-black sea resolves from owl into a little skin-skeletal shape lying horizontally with a pounding triangle in its centre, its weeny bone legs bent up to its chest.

'Can you see the hands?' Rachel asks as she highlights and zooms in. Not really, just a tiny, sleep-twitching fist covering the head in a boxing defence. Then it moves, shifting around onto its side as if trying to get comfortable in bed. This is what we will both remember: staring goofily at its incessant wriggling, a tiny quicksilver ghost messing up the bedclothes.

Along the hospital road, the white and pink cherry blossom is coming. There are spears of snowdrops in the gardens. I didn't see them on the way there. The world smells cleaner. I confess to Rosie about seeing the owl in the ultrasound and she smiles. Later she says, 'Why don't we go down to the edge-land again? I'd like to hear the owls tonight.'

I am learning to value the tiny subtleties in times of day. The edge-land changes depending on when you arrive, but also how you arrive, almost *who* you are when you come here. The evening feels celebratory. The sun tints bare trees into fountains of gold, turning lone crows in their crowns into weathervanes, beaks spun round to the south-west. That long blurry fortnight of snow was blinding; now it's as though I've have had my eyes tested and been prescribed stronger glasses. The skin of the earth looks grazed in patches, like a toddler's knee; shoots of wheat push to the surface in a thin wash of green blood. Elder buds in the holloway are cracking open with tightly folded clusters of crimped, red-edged leaves. We walk quietly to the wood down a passageway brimming with momentum and trembling air. Rabbits have been digging new holes in the thawing ground, dredging deep terracotta earth and scattering it under a hazel. Silver catkins soften the dark stems of sallow. It isn't even dark yet, but the male tawny owl is already broadcasting on long wave, calling high so his notes carry across the weir.

'*Listen!*' We hiss under our breaths, grabbing each other's sleeves. He is less than ten feet away on the low branches of a pine, facing northwards, away from us.

You hear people talk about having 'our song', a tune with significance that they've claimed as their own. Something that became meaningful by chance, by coincidence. If it was the same with birds, ours would be an owl. A long-eared owl sang us home from one of our first dates

– a night walk from pub to pub in the Lake District; resident male tawnies provided moonlight sonatas in the communal grounds of our first flat in London. Most impressive, though, was the snowy owl that gatecrashed our honeymoon. Taking a remote Cornish cottage in January for a few days seemed a sure way to get privacy, but one morning we looked through frosted windows to see the narrow lanes and fields choked with cars, binocular-browed twitchers and news crews. They had descended overnight from all over the country in the hope of catching sight of a rare visitor, a nomadic bird of the high Arctic that isn't usually seen so far south. That white phantom haunted the copses and fields of the Zennor coast 'like a decent-sized lamb in the trees', as one local put it. But despite our best efforts and enviable position, we never saw her. Then one morning, ironically, as a curtain of snow approached from the west, owl and people vanished.

To get a better view tonight, we creep over pine needles and dead leaves but the tawny bores of our amateur stealth, turns and fixes us with a black-marble gaze. It feels like being caught by your parents secreting a girl into your bedroom. His ragged, dark-ringed face is fearless, frowning; his shaggy cryptic plumage blends perfectly with the bark of the tree. Tawnies roost by day but they love sunbathing and the warmth of evening light bronzes his feathers. We aren't the only ones who've seen him. A great tit keeps its distance but performs elaborate flicks of wings and tail, firing rapid-note alarm calls. Daylight exposure is risky business for owls. The characteristic head and body shape is unique and, once recognised, mobbed. In times past, hunters covered trees with birdlime, stationing an owl decoy so that other birds would fly at it and became stuck. This is what the great tit's dance moves are about, to marshal others to pester and pick at the predator until it moves away. I've seen this only once before in a woodland in Kent; mistle thrushes appeared and swarmed the top of a foliage-covered oak, as if grassing up the concealed female tawny within – *She's here, she's here*. Then came wave after wave of their sharp-beaked, full-frontal attacks. Few birds are so hated and feared by other birds. Even hawks are better tolerated. As Roman poet Ovid wrote of an owl: 'She is a bird indeed ... but conceals her shame in the darkness; and by all the birds she is expelled entirely from the sky.'

Back before owls became a fixture in our fashions and furnishings, they weren't much regarded by us, either. From biblical times, they were considered a winged harbinger of ill and the sign of darkness and ruin. In the words of Pliny the Elder, writing over 1,900 years ago: 'He is the very monster of the night ... if he be seen, it is not for good, but prognosticates some fearful misfortune.' This reputation probably stems from the owl's tortured-sounding call and predilection for haunting ruined, abandoned tracts and buildings. But mud sticks. In the sixteenth century, Shakespeare reflected similar English folk beliefs in *Macbeth*: 'It was the owl that shrieked, the fatal bellman, which gives the stern'st goodnight.' Confusingly, though, in the eighteenth and nineteenth centuries many communities and regions believed the opposite – that the owl was as much a forecaster of good and success as it was of evil. Hooting was said to foretell the birth of a girl; an appearance near the house of a pregnant woman was said to ensure an easy delivery or the birth of a boy. Here in Yorkshire it was once claimed that whooping cough was cured by a good bowl of 'owl broth' and numerous recipes suggest the bird's eggs should be eaten for rude health. Farmers certainly appreciated their skills in keeping field and wood mice down. And there was a belief among certain Native Americans that the cry of the owl was its mournful remembrance of a golden age when men and nature lived in harmony. I like that tale the most – owl as balladeer forever lamenting the Garden of Eden.

But enough of this. It's too much to carry on little shoulders. The tawny has no interest in our speculation and superstition. It is what it is. With a last look down, he tilts his face up to the dying light and then folds away, passing low and fast over the river.

March marches on. Rosie grows. Leaves fur the trees, save for one giant sycamore in the wood that is reluctant to join the party. I can follow its naked shape from the faint feathery top down to its exposed roots. Doing so is akin to tracing the course of a mighty river on a satellite photograph; every branch is a tributary, each bursting bud at twig's end a rising spring. It could be the Nidd, frozen, black and hauled upright. Nearby, about thirty metres from the weir, is an old beech where I'm sure the owls must be nesting, high up in a rotted hole between two great branches. The entrance is well concealed with ivy but if I listen long enough in the dark the male appears to circle from this spot in long, extending sorties. Borders are policed by his regular screeching checks that make me jump when near and at the limit of their territory become the soft, whistling high notes of wind in a chimney flue.

The male is increasingly hunting by day now, proffering gifts of what I assume must be voles, rabbits and mice to his brooding mate. I've bought a pair of binoculars to carry with me and raided my mother's old birding guides for more information. They recommend determining diet by searching for the owl's 'pellets', the disgorged bone and fur of its prey that act as a kind of food diary. It's easier said than done and an effort to find anything in the decaying ground layer of last year's leaves and emerging forests of dog's mercury, wood anemone, lesser celandine and ramsons. Even so, the searching is an education. Winter still holds the edge-land's hollows but everywhere else incidental riches grow, new life through the garlic-smelling earth. There are carpets of bonsai-like nettles and cleavers growing in perfect miniature. Marsh marigolds' golden cups cluster in the streams and little waterlogged culverts near the river. I have a five-minute staring competition with a young wood mouse as it emerges from under a rock two metres away. It is unperturbed until I reach for my phone to take a photograph. I must be the first human it has seen.

It is about 5 p.m. judging by the rush-hour rumble and horns on the roads from the south and, across the river, the boom and clatter of metal as arable machinery trundles back to barns. Shadows are lengthening and there's the rusty *cook-cook* of a pheasant and twittering robins. The male tawny is alone on the edge of the wood, perched on the wrist of an oak's long, thick limb. There's enough mustard light to use the binoculars and, in the natural hide of the holloway, I prop my elbows on a low branch and take in the details of his muted and mottled cream and chocolate feathers. They have an unbelievable smoothness, like a luxury truffle, the sort you buy in those twirling seashell shapes. He moves his head mechanically, rotating that facial disc like a satellite.

Actually, the more you read about a tawny owl's hearing, the more sci-fi it all feels. Their ears are roughly ten times more sensitive than ours and able to detect low-frequency nuanced sounds over considerable distance, even those as subtle as the rustlings of prey moving through vegetation. I say 'ears', but they're actually two openings positioned asymmetrically either side of the face hidden by feathers that have adapted to be 'transparent' to sound waves. The left opening is positioned higher than the right, which points downward to improve the owl's sensitivity to noises from below. All sound coming from beneath its line of sight is therefore louder in its right ear. Running through the skull and linking the two eardrums is a passage packed with auditory neurons that tell the brain the fractional differences in the times of arrival of sounds to each ear. Instantly translating this data – *right, left, up, down* – the owl precisely maps and pinpoints the source.

Then he falls off the branch like he's been shot and I lose him in my glasses. Lowering them and squinting, I catch a brown blur plummeting towards the field and rush to fix them on him again. It's all so swift and silent; his wing feathers serrated like a comb to eliminate any flying noise. His head stays perfectly still, his spectacular vision is now locked-on but, mid-flight, perhaps only five feet from the ground, he makes a correction and banks left, bringing his feet forward and extending his talons, swooping his wings down as though he means to smother the prey for good measure. The moment of the strike is lost in the dwarf wheat but I watch as he immediately bobs his head in a series of flinging, violent blows. My heart is in my throat. Then, after a minute, the owl looks around nonchalantly and rises, carrying off the crumpled grey ball of a baby rabbit. A life ended before it has begun; a stomach fed. All the while the pheasant and robins continue with their idle chit-chatting. Nothing sees; nothing cares. The hunt, the death, all of it seems so shockingly routine.

*

Later I come home to find Rosie opening a letter from the hospital confirming the date for our twenty-week ultrasound scan. As she jots the date in the calendar, I read the much-photocopied NHS literature explaining how it all works, how the probe sends high-frequency sound (1 to 5 megahertz) into the body and listens to the echoes that return. Apparently the sound waves hit the boundaries between fluid and flesh or muscle and bone. Some bounce back to the probe; others travel further to be reflected by another boundary. Perched on the belly, the probe hears the millionths-of-a-second frequency differences between them and using the equation of the speed of sound in tissue (1,540 metres per second) the machine processes and resolves the distances, mapping the body's internal landscape in a two-dimensional image. The leaflet also carries a word or two of caution. The more detailed investigation is also known as the *anomaly scan*. You find out if all those boundaries, bones and bits of flesh are behaving properly. 'The ultrasound occasionally detects some serious abnormalities,' it warns at the bottom, 'so you should be prepared for that information.'

Again, so routine.

The anatomical and emotional changes in women during pregnancy are well known and documented, but impending parenthood alters a man too. There's a lot less said about that, but I'm feeling it already. The skin becoming thinner and more sensitive, the world even more glorious and cruel.

The week is crashing out with the vindictiveness of an army laying waste to the ground. A wet storm-wind shrieks from the west into the edge-land across miles of hill and farmland. It tells of wars being fought and battles approaching. 'This is unbearable,' shouts a lady buffeted from car to front door as her umbrella leaps from her grasp and vanishes over a hedge.

The gales are strong enough to drive leaves from the wood all the way up Bilton Lane like scuttling refugees heading for the safety of town. I head in the opposite direction, passing the crossing point before slipping left over the little ford and up the lane. There is nobody else here and the trees usher me away with frantic gestures. *The wrong way! You're going the wrong way!* A swirling acrid wind makes it hard to breathe and look in any direction for long and I envy the rabbits and voles huddled now in burrows. A coal tit crashes into a holly looking for protection; I seek shelter too, running for the holloway as shouts and moans howl through its hazel walls. By the wood an ash flails and I wrap my arms around its trunk to stay upright. The whole place is in distress. Canopies throb and merge as pines roar their anguish over the river. To escape the barrage of debris and air I wedge myself between two pine stumps. It is at once terrifying and invigorating, as if I am looking up from the ocean floor as a tsunami passes overhead. But if I kneel I can just about see the owl's tree and follow its swaying trunk with the binoculars. Chances are the female tawny is inside nesting. It worries me. Wind and rain smashing and percolating through woodland destroys an owl's ability to use sound, making hunting extremely difficult, and the pair must be hungry. Regardless, the field guides confirm she will be incubating between two and five white eggs, each laid a week apart. Weather or no weather, life goes on.

The wind pitches louder, higher like a dynamo being wound into a scream and a trunk falls somewhere. A quick look over the parapet; it's not the owls' tree, but how long until it is? In the darkening underworld of the forest floor I feel the thin line between life and death running like a fault below me. Still, I stay put out of some sense of fatherly solidarity. *It's just nature, though*, I tell myself, *whatever happens*. Then, from nowhere, a rising tide of emotion hits me and I find I'm thinking of my friend Peter who died unexpectedly, unnecessarily, in an accident seven months ago. I feel the same slow soul-sadness I've experienced regularly since answering the 2 a.m. call from his cousin, an unreal mix of helplessness and incomprehension. Then that yawning absence. Like touching an electric fence, it's the sudden jolt of finding yourself at the margins where human and nature step outside the neat categories we give them, those moments when we're suddenly forced to confront the non-negotiable fine print of our existence. These are the other moments when we hear the undetectable frequencies and unforgettable sounds, like on the night I heard about Peter: the crack in his cousin's voice and, after I'd told her, Rosie's repeated '*Oh no*', a trembling hand half-covering her mouth.

Out here in the edge-land, transformation is in everything. With a million life-and death-moments happening each millisecond, you never escape the sense of shifting states. I know the water I can hear crashing over the weir is destined to become open sea and although it will no longer be 'river', it will still exist. And one day it will be resurrected into rain again. It's just the same with the baby rabbit and the owl. Energy never runs out, it only changes from one form to another. It can't be created or destroyed. Every joule hurtling past me was present at the beginning and will be there at the end. There's consolation in that, for sure, but try telling it to the dad pacing a hospital corridor as his baby is operated on for congenital heart failure. This is the problem with 'nature': it is ambivalent to what makes humans tick. And yet it *is* what makes us tick. The two can be hard to reconcile sometimes.

For three nights I have vivid dreams of being in a crow's nest on a raging sea. Our house crashes, woos and roars as the air screams around it, like some Victorian machine pushed to its limit. But now the pangs and pings outside sound like the engine is finally cooling and contracting.

As soon as I can, I run down to the wood, dodging the broken sticks, plastic bags and fast-food cartons still skidding under cars and over pavements. Garden conifers suddenly strain and roar. A screeching call of distress sounds from a garden by the crossing point, like a blackbird trapped by a cat's claw. Hurrying to the gate I see a car in a drive with its bonnet up, a man wedged sideways underneath in mucky overalls. He catches my look of surprise before turning back to the real source of the noise, a faulty ignition.

From the old railway northward the earth is battle-scarred. Cracked trees slump as if contemplating the bone-white heartwood bursting from their bodies. The air has ripped through the hedges along the lane, slashing and tearing as it scattered felled limbs, snapped sticks and shredded young foliage everywhere. Another fizzing click-ing sound cuts the air, this time from over the fields. It is an electricity pylon crackling from its high triangular crown like a giant insect scratching its mandibles. In a dip in the farmland to the north, a copse grapples with the vanishing wind in a final skirmish. Then the earth falls quiet again. I stand below the pylon listening to its alien song as evening drifts over the wounded land, softly, a shaken sheet floating down onto a bed. I'm wondering how anything can have lived through it when the tumbling notes of blackbirds pour from the holloway and wood. Then, louder, flung up towards the milk-moon above, the long *hu . . . hu-hu-lo-hooooo* of a tawny owl.

Fast forward. We are back in the Antenatal Clinic. Once again we take a seat on the line of smart red chairs. This time our hands are spread over Rosie's swollen bump. Our disinfected palms and fingers feel the fluttering kicks and movements from within. Opposite me, the owl is still Blu-tacked to its wall, solitary in that broccoli tree stuck between the noticeboards. How long it'll be there, though, I'm not sure; a toddler turned loose by his exasperated and heavily pregnant mum is having a good stab at tearing it down. 'Me-me,' he says, dragging at its corner. *You said it, kiddo.*

Among the many fantastic shapes and colours that make up the bird world, tawny owls do look remarkably similar to us – the two-legged vertical posture, the rounded head, the big eyes, the binocular vision, the cheek-like face and that beak projecting downward like a little hooked nose. Even the white-grey tufts of down around its face could pass for an unkempt moustache on an old general. These resemblances, of course, explain why they've been elevated into the realm of soft toys and grace our furnishings, clothes, bags and wallpaper. It's vanity. We're rewarding their human cuteness. The subconscious connotation is an old one, the Disney view of nature as something we can master, adjust and appropriate. Something just like us. In many instances this may be true, but watching birds of prey up close or the proximity of a birth or death forces us to think differently: we engage with nature directly, emotionally, microscopically, at the level of atoms, cells, blood, flesh and bone. We briefly see under its skin and witness its contrariness, its randomness, relentlessness, ruthlessness and beauty.

Sitting with my hand hovering over my unborn child, I feel none of the indestructibility that gives our species its hard edge and cynicism. Rather, I'm acutely aware that we are as susceptible to fickle fortune as every other living thing. In a heartbeat or baby-kick, all of our modern disconnection can be stripped back to its opposite. We seek reassurance; I hold Rosie's hand. And there it is: the hand, the arm around the shoulder, the soft kiss on the forehead, the inner voice you find uttering a silent prayer in a hospital waiting room. We often forget, but this is nature too.

'Now, would you like to find out the baby's sex today?' Rachel asks, closing the blind as Rosie lies back and rolls up her top.

'No, I think we'll wait. We'd like a surprise.'

'Good,' says Rachel. 'I think that's nice.'

Lights. The dashboard fires up. Another flight, this time over familiar ground. The probe circles, perches and listens. Rachel talks throughout this time, listing body parts as though reading the shipping forecast – *brain: good; arms: good; shoulders, ribs and pelvis: good.* She pauses occasionally to measure or make notes, and then turns the monitor around. Everything is discernible now in that black-silver sea: fingers, feet, hips, a little nose, knees. I can see its body flex and reform in amazing detail as the noise of the heartbeat pounds fast and strong, like thundering hooves.

'You have a very healthy happy little baby,' Rachel says eventually. 'Would you like a photo?'

I fold the printout into my wallet, a cocoon for the little one.

*

But that's all to come. Tonight I go down to the wood again. The tawny owls' nest has survived the storms, for the male is out hunting early again, bringing food to his mate as she incubates their eggs. Setting up my binoculars on the pine stumps, I try to picture the embryonic owlets tucked up there in the tree and think about their chances. It wouldn't be wise to get any closer, though; tawnies are famously vicious in their defence of the nest and known to attack dogs and humans if they stray too near. Such protective spirit is a fairly short-lived instinct, however. Their parents will care for the young birds for two or three months after they fledge then, around August, the juveniles will disperse to find a place to call their own. If they fail to find a vacant territory, they'll quickly starve or, in weakened states, fall victim to predators. Nature will take its course. No hospitals, medicines, warmth or love to intervene.

I leave them to it and walk back over the crossing point, heading for town to buy groceries for dinner. I carry that call with me and think about how lucky I am. As Edward Thomas has it in his poem 'The Owl':

> *All of the night was quite barred out except*
> *An owl's cry, a most melancholy cry*
> *Shaken out long and clear upon the hill,*
> *No merry note, nor cause of merriment,*
> *But one telling me plain what I escaped*
> *And others could not, that night, as in I went.*

THE UNION OF OPPOSITES

I

The day of the spring equinox is marked by a last, late, unexpected rush of snow falling soft and heavy under the cover of night. Nothing alters human consciousness so entirely and immediately as waking up to a different land outside the window. 'A change in the weather,' wrote Marcel Proust, 'is sufficient to recreate the world and ourselves.' And he was right. This morning feels strangely hallucinogenic, like some prankster has spiked the municipal water supply. By nine o'clock everybody's attempts at getting to work have been abandoned. Men lean into wheel-spinning cars helping return them to garages. Neighbours who've barely exchanged two words before lock arms and creep along Bilton Lane, their faces fixed in smiles. The talk is of schools closing, train cancellations, road closures. A day off.

Parents drag sledges over a crisp white countenance, carrying muffled cargos of toddlers down towards the edge-land. But there is no edge-land today. Rural and urban have become briefly inseparable regions. There is only white earth and a dish-rag sky drifting with a few final flakes, like the aftermath of a pillow fight. Garden, yard, fence, farm, road, field – all are buried. Psychological barriers similarly forgotten, strangers surge together over the old railway and into the meadow, unaware of exactly where they are. It doesn't matter; all the earth is theirs. Children scream at the sight of space and wade into snow, eager as sailors reaching the shallows of shore. Dogs bound after them, barking and biting at scalloped waves. Unsteadied, they are oddly fawn-like. Robins and chaffinches strike up from hedges and a wind blows through the belts of distant pines, twisting their tops so that they appear desperate to look. Birds, trees and sky congregating for a spectacle: the people returning to the land. This is common ground again.

The scene that unfolds is dreamlike, out of time, refusing to fade even as the day yawns and reaches for its glasses. Snow brings an ecstatic calm, the same high, lightheaded buoyancy you feel after crying. Although cold, the air is clear and mistle thrushes burble from the wood's edges. Irrepressible, nimble-footed, life dances over the fields in the delicate notes of birds and a dazzling searchlight sun. Beneath drifts, lolling green tongues of lords-and-ladies – *Arum maculatum* – lick away at the snow; young nettles, Jack-by-the-hedge and soft swords of grass tickle this crystal skin. I crouch down against an oak tree at the meadow edge and let the sun's rays fall across my face, sending me into a pleasurable dizziness, a half-trance. This is the hour, the glimmering union of seasons. Everything testifies to it.

Judging from its width, my guess would be that the oak is around 250 years old. Easily the oldest I've found here. A survivor. For a moment I wonder at how many red squirrel drays and wryneck nests it's held across the vanished centuries. Maybe wildcats scratched claws down here among its adolescent roots. Today its choirs are in full voice: wrens trill, there are the bicycle-wheel squeaks of great tits and everywhere pours with the thick flute-notes of blackbirds, like blood flooding back into numb limbs. The exodus from town is unstoppable and the meadow throngs with people. Hands red raw, children roll up the snow in creaking channels to make the bodies of snowmen, leaving deep trenches behind them. Their brightly coloured coats and hats bob as they squeal and dip in and out of their burrows. I squint into the glare and imagine time slipping back to a late-August afternoon when the oak at my back was an acorn. Instead of the snow, I'm watching families harvest the plaited-hair spikes of silver-gold wheat. Boys and girls are tracing ancient patterns over these fields. We've all been here before, sweaty, bent and hacking with hand scythe and sickle, cutting callous-forming avenues through whispering stems, reaping, rolling and stacking sheaf. For a moment I'm part of another union, a brief and rare return to the earth for us landless masses.

Perhaps those old enough to remember might carry this moment with them through largely room-bound futures. Dispersed into offices miles from here, the dim recollection of it may manifest sometimes in a bright wash of spring sun falling accidentally across a desk. Pausing, eyes will look out through the glass and feel a sense of belonging to a larger world. Or when old and alone in a nursing home, a window opened a crack for air may fill the room with the smell of swelling life and fresh snow. That aged child will sense the union of seasons again, close their eyes and escape through the gap.

Such golden moments pass, evaporating into air, melting into earth, leaving only that which we carry with us. It's midday and already the faint watermark of a

moon is imprinted in the milky blue to the east. Children
are rounded up, brushed down and carried back along
the old railway, placated with promises of chocolate and
TV. They cease struggling and slump dull-eyed on shoulders. Time's up. The drug is wearing off. Conversations
dwindle. Laughter ceases. I see why. Combined with the
heat of the sun, their relentless industry has revealed divisions again. The gardens, fields and roads are exposed;
snow has melted off the walls and fences and been
stacked into gritty piles; enmeshed wires and concrete
stanchions are uncovered. This is mine; that is yours. Shut
the gates and doors. Slowly the town reabsorbs its
populus, sorting and separating down freshly shovelled
driveways, sluicing from minds this morning of freedom,
turning thoughts to urgent bills, unanswered emails and
phone calls. And I can feel this gravity too but I resist and
remain, here beneath the oak. The day has something
more to show.

Later, I'm walking with a wilful sense of trespass
through the wood and up into the fields. There is that
same smell as when you rub dock leaves on bare legs. The
sun has faded large tracts of snow; the land is blotched by
wet wheat gleaming with an unearthly light. Out of one
patch, a hare emerges. At first I assume it's a rabbit until
it rises tall, sniffs and hoists those long, Indian-ink-dipped
ears towards the sky. Framed by the chequerboard town
and purple-white of distant moors, I watch it hop in a
circle then fling itself towards the field edge, through
a blackthorn and out of sight. Most memorable is its coat:
the gingery dun of high summer soil still brushed with
the grey flecks of midwinter camouflage. The hare, the
common hare: a hybrid, a union of opposites running
through the edge-land.

The man who encounters the hare
Will never get the better of him,
Except if he lay down on the ground
The weapon he bears in his hand
Be it hunting-staff or bow
And bless him with his elbow.
And with sincere devotion
Utter this one prayer
In praise of the hare.

Winter and summer. In the wheel of the year, these are the most distinct seasons, the opposites. Everything about them is contrast: darkness and light, cold and heat, life and death. Winter even looks like the photographic negative of a summer day. Where these meet in the spring and autumn come days of union. You find flashes of both extremes. Falling snow becomes foaming blossom; rain is suddenly refracted by bright sun; morning frost fades into water-colour evenings of peach and pale yellows. Balance is everywhere, and not in some wishy-washy sense. I'm talking planetary equilibrium.

The word 'equinox' derives from the Latin *aequus* (equal) and *nox* (night) because in the northern hemisphere these are the moments in March and September when day and night are roughly the same length. Briefly the sun sits exactly over the equator and the tilt of our planet's axis is neither away from nor towards it. The north and south poles stick straight up and down, like a cocktail stick through an olive.

We live in a world where the shift from dark to light is nothing remarkable. It is a constant process. The switching on and off of electricity falls to fingers countless times a day and we conjure heat at the touch of a thermostat.

At hundreds of miles per hour 35,000 feet up we move between hemispheres, passing from winter to summer in minutes. It makes it hard to conceive of the impact the spring equinox held in earlier times. To eyes long tired of merciless winter, a strengthening sun returning to banish the darkness was a supremely powerful moment. As the icy crust split, destruction gave way to re-creation. Land was miraculously fertile, fecund and full of energy. The equinox saw the arrival of a resurrecting light and, to a species dependent on earth and animal, it simply meant survival. Understandably we sought symbols that embodied this sequence and associations flowed and formed. As is our habit, we needed to identify and name the unknown, to give it shapes and representatives. Among all the candidates one creature was a front-runner. Widespread across the northern hemisphere and thriving with the birth of farming, the hare leaped into our consciousness and imagination.

For an animal that spends its existence above ground in all weathers, the 'brown' or 'common' hare (*Lepus europaeus*) remains a mysterious and perplexing thing. Usually nocturnal, solitary and almost always in hiding, it is practically invisible most of the year. In winter it seeks refuge in woodland, hedges or its well-concealed 'forms' – the shallow scrapes it makes in soil. By early May its activities are lost again among the crops and grasses of arable fields. As such, it appears to belong entirely to the spring, materialising from the substratum and congregating to play out energetic breeding games. Promiscuous and highly productive, it is new life in fur form.

In several languages the hare is, literally, 'the leaper', one who springs up, with the connotation extending to the dawning of being, as much as a new season. In Egyptian hieroglyph the symbol of a hare denotes 'to be'. It is persistence made flesh. The name of the god *Michabo*, creator and personification of the sun, regarded as a common ancestor across all Native American people, is a compounded version of *michi*, meaning 'great', and *wabos*, meaning 'white hare'. *Michabo* – the Great White Hare, God of the Dawn.

Closer to home, over the same parcels of earth I've known at sunrise, the Anglo-Saxons worshipped *Eostre*, the Germanic Goddess of Dawn. Her name changes depending on location – Eostre is Northumbrian dialect; Eastre, the Old English of the West Saxons; Ostara is Old German – but all are thought to derive from the ancient dawn goddesses Eos (Greek), Aurora (Roman) and Ushas (Indian). The eighth-century English monk Bede notes in *De Temporum Ratione* that before Christianity absorbed Eostre's name and rituals for its own celebrations of resurrection, April was 'Eostre-monath', the dawn month. Some claim Eostre had the head and shoulders of a hare, even that the animal was her attendant spirit, carrying a procession of lights or torches into the world. Others believe this is merely revivalist neopaganism, popularised in the twentieth century. Whatever the truth may be, the evidence of the hare as an early and enduring cultish figure is testified to by its Christian associations with witchcraft. Its demonisation, and later appropriation, by a newer religion tells of an existing position of reverence in people's minds. Certainly by the time of conversion this symbol was too engrained to be lost; it's tempting to imagine the hare was simply baptised and born again, neutered, turned to chocolate and re-imagined as our harmless Easter Bunny.

There is also a connection between hares and that other Easter tradition, the giving of eggs, a custom that dates back at least a thousand years. Brightly coloured and

reputedly delicious, the eggs of the lapwing were traditionally collected by adults and children and eaten around
April, becoming so popular throughout the Victorian
period that preservation orders became law in 1926 to
protect the bird. This precedent to the modern Easter egg
hunt was made easier by the fact that the bird nests on the
ground. Much like hares, lapwings prefer fields for their
large range of vision, vital for detecting approaching predators. So similar are their habitats that hares have even been
known to borrow lapwing nests to hide in. People disturb-
ing a hare might have found a nest of eggs beneath. It's
understandable that they would leap to the conclusion it
was the hare that had laid them.

Despite being a creature of the field, the hare springs
up as the hero in many flood myths, including folk tellings
of Noah's Ark. The flood itself is, of course, a metaphor
for new beginnings, symbolising the world's resurrection
from death – the stormy darkness giving way to renewing
light. In some narratives, the Devil tries to scupper the Ark
by creating a succession of holes in its hull. Noah plugs
each one until, running out of suitable materials, he cuts
off the female hare's tail to use. In another version, it's the
hare's foot that Noah uses to plug the leak, driving away
the Devil in the process. The superstition of carrying a
hare or rabbit's foot for luck persists, but in the story the
act kills the animal. God repays its sacrifice by granting
the remaining hare on the Ark – the male – the power to
give birth.

Until as late as the nineteenth century in Britain, this
wouldn't have seemed as ludicrous as it might today since
it was widely believed that hares were hermaphrodites.
Since the time of the Greeks the animal was said to have
the ability to change between male and female from one

month to the next. Pliny, Philostratus and Plutarch all ascribed to it. Old Welsh law accounts for it in its assessment of animal worth, specifying how 'for one month it becomes male, and the other female'. This duality was likely a legacy of earlier cultures and religions that regarded hares as being both masculine and feminine, an embodiment of the equinox's moment of balance, the union from which creation breaks forth into life. Myths of mutating sex are also probably down to the hare's physical similarities. Males and females are almost indistinct and, despite what we thought until relatively recently, the females box away the advances of prospective mates. If you believed, as most people did, that it was solely males boxing each other for the right to mate, it would be easy to mistake female for male and vice versa.

Over time an even wilder fertility myth took hold: that the hare was capable of self-impregnation. Through association, rabbits too were said to possess the 'gift' of reproducing asexually. This reputation for virgin birth saw their stock rise in Christian symbolism, explaining why you find hares and rabbits cropping up in sixteenth-century religious paintings. Although incapable of immaculate conception, the workings of the hare's reproductive system still fascinate. Recent scientific studies claim to prove that the hare is capable of 'superfeotation', the ability to become pregnant a second time while carrying a first batch of young. In other words, two eggs from two different ovarian cycles can develop simultaneously in the same body. But not all scientists are convinced. The debate rages; academic fur flies. One thing is sure, however: the hare as mythical life-bringer and fertility icon remains irresistible. No matter how fleet-footed it may be, Britain's fastest mammal can't outrun its past.

The hare, the hare-kin,
Old big-bum, Old Bouchart,
The hare-ling, the frisky one,
The way-beater, the white-spotted one,
The lurker in ditches, the long eared,
The slink away, the nibbler,
The scutter, the fellow in the dew
The animal in the bracken, the springer,
The swift-as-wind, the skulker,
The shagger . . .

It comes to me out of the blue. A memory. It must have been fifteen years ago. Further west and north of here in hill-farming country I was out walking on a day so hot the horizon trembled. I had stopped for a drink near a field where an old hand was cutting hay. Seeing me, he swung down from a tractor cab, tanned as a cornhusk, and leaving his rusting Massey Ferguson idling, lit up and sauntered over. I'd just watched a leveret – a young hare – sprung by his mowing. It darted silkily from the approaching storm of seed, metal and machine, zipped over the cut stubble and plunged into the fog of grass again. I told him as much.

'Aye,' he said. 'Good when she runs.'

She, I remember that. The hare was feminine.

'Don't they always run?'

He shook his head. 'Hit two this year s'fa. Terrible.'

We sat for a while as he worked through the last of a crumpled pack of Silk Cut. He smelled of hay and earth and engine oil. Whether true or not, he was adamant that male hares – he called them 'Jacks' – ranged great distances in their pursuit of mates, but that 'Jills', the females, remained in the area they were born. A 'matter of fields', as he put it. It was clear that even an accidental killing made him uncomfortable. There was an unspoken reverence at the edge of his words, a sense that the hare was something other, a creature with knowledge that runs deep.

'Jills know lay a' land,' he said. 'Every rut, blade and furrow.' He dropped his dog-end and dug it into the earth with the heel of a boot. 'Know it better than owt else. E'en old buggers like me.'

I'm thinking of his words as I again cross over into the edge-land. It's getting on for mid-afternoon but with the sun out for a full twelve hours each day now there's time enough to catch a hare. I've a notion of witnessing that moment again, that bubbling up from the earth. With thoughts of new life running through my mind, I want to see the myth made flesh, the old magic up close.

The windows of the houses along the old railway flash with light. A dog barks without break. There is the growl of traffic from the Ripon Road. Horns are unleashed and angry air brakes hiss at each other. Ringing through the housing estates' roads and cul-de-sacs is the twisted metallic melody of an ice-cream van, a sinister-sounding 'You Are My Sunshine' in a tortured staccato. It's strange to think hares would tolerate this and remain on ground less than half a mile from the encroaching urban sprawl. But they do. This is their turf. Like the farmer said: *Every rut, blade and furrow.*

Bilton Beck's wooded gully is clogged with farm sacks, branches and litter, but here the engine noise and Mr Whippy jingles become lost in the sound of partially obstructed water channels and air-blown beeches. Shadows of larch pitch over fallen walls now devoid of purpose, velveted with moss. This rift demarks another empire. Just east of its treeline, close to where I saw the hare on the equinox, I sit on the muddy edge of a field. It seems a good place. Here the ground curves away in rows of spring crops, neat green tramlines interspersed with sandy belts of earth. They ripple up to a long bank at the field's top where a scrappy hedge of hawthorn, blackthorn and hazel hides a stray gorse flecked with yellow, coconut-scented flowers. Up close, between my clay-clodded boots, the wheat looks like cosmetic hair plugs sprouting in the precise rows left by a mechanical seed dispenser. Each stem is barely ten centimetres tall. High enough, though, to hide a hare.

Actually there are two, fifty metres apart, hunkered either side of a telegraph pole. Each time the breeze riffles the crops I pick out what appear to be the tops of stones or mounds of earth turned by a plough. Through binoculars they resolve into haunches of folded leg or pelvic bone felted with a fur the colour of the underlying soil. Then they're gone again, into the green sea. It's what Freud termed *unheimliche* – uncanny – this sitting here watching the endless becoming and vanishing in the mud and shoots. Occasionally one of them – a Jack, I presume – lifts itself onto legs that are too gangly for walking. He turns in a small circle and drops back into his form. A hare's movement seems plagued by the flicks and judders of restrained energy, as if carrying an ache that can only be relieved by running. The rest of the time it's as though they're absorbing the earth's energy, tapped into a ley line, shivering with pent-up static.

At 5.20 p.m. exactly, the angles of the field and descending sun form lakes of light that spread across the land's surface. A secret message is communicated. On heat, the Jill releases pheromones, which act as a starting pistol for the Jack. He rises, periscopes his head and then whips towards her. Watch a hare run and you see how closely they resemble deer taking flight. An old English name, *stobhert*, means 'stag of the stubble' and it is well earned. Ears folded back into the long cone of a speed-cyclist's helmet, their acceleration is extraordinary; hares can reach thirty-five miles per hour flat out but she is ready for him and rebuffs his advances. They roll wildly for a moment and then she is free. Somehow she materialises further up the field, covering a distance of twenty metres or so in a second. He catches her and they circle. This time they come together on hind legs in a flurry of boxing blows, punches and flailing paws. Soon it looks more dance than fight, like they're holding hands and spinning the other around, the way kids do in a playground. Tumbling and chasing they cover the width of the field three times, then just when I think she's tested his mettle enough, they both stop dead. Ears prick up. Eyes stare. Bulging, yellow, goat eyes. Sex is suddenly off the cards; threats take priority and there's trouble brewing. I can't discern the cause, but both are spooked by something and scatter north, sinking into the wheat sea, ears flat along their bodies, heads down in the earth. At the point they disappear they are ten metres apart. That's my guess anyway. With a hare you can never be sure.

I wait a while longer with the sun on my face but know they've already crawled back to their imprints, pushing their bodies into the soil, fusing with the land. A pair of buzzards circle over the holloway calling in shrill two-tone cries. A final whistle. Crows float up to mob them and I move too, heading back the way I came down the gully and along the trickling beck, turning east where it widens to meet the Nidd. Wild garlic grows in ranks of massed spears here and I fill my coat pockets with the stuff. Around the trees near the old weir a stronger smell drifts. Sharp and smoky. I know it before I see it. In the middle of the strewn foundations of the vanished mill a fresh campfire smoulders with half-burned green ash branches. There's no one about, but it's unsettling to discover it in the same place I found the fire remains at the beginning of the year, positioned specifically to look over the river. It suggests a more regular habitation. The Nidd splits over the weir like Brylcreem-parted black hair, then cascades onward noisily, cloaking all other sound. I couldn't hear if someone was returning until we stood face to face with each other. So I leave. Quickly. Amid a thick carpet of dog mercury, I step over plastic bags filled with tins of ravioli and vegetables. A threadbare rucksack is stashed by an alder. Squirming out of its top are the crumpled brown folds of a sleeping bag.

> ...*the squatter in the hedge,*
> *The friendless one, the cat of the wood,*
> *The short animal, the lurker,*
> *The fidgety-footed one, the sitter on the ground,*
> *The sitter on its form, the hopper in the grass,*
> *The get-up-quickly,*
> *The one who makes you shudder,*
> *The animal that no one dare name.*

The smallest bedroom at the back of our house has been dark every morning since we moved here. Now with the sun up early in the east and high over the roofs by 7 a.m., the first thing I see en route to brewing coffee is a sunbeam illuminating the stairwell. Bleary-eyed, I push open the crack in the door and enter a room of light. Normally it wouldn't register but for two things: this is the room we're decorating to be the baby's nursery and on its wall is a foot-high sticker, a Beatrix Potter rabbit sitting side-on to face the rising sun. I don't see a rabbit today, though; I see a hare.

Early morning and dusk are the best times to be out. It's when they feed. I take the long way round, avoiding the wood and its incumbent, and set up a position under the grey of morning. My eyes are drawn to the treeline and an illusion of more campfire smoke, but it's just wisps of mist. Dew hangs off the dandelions. The haze slowly lifts. I take root and sense the air of loneliness that hangs over all modern arable land, the absence of people exaggerated by the remnants of unremembered existences, the broken bits of pottery and rusted horseshoe fragments poking from the soil. Away to the south the 9V-battery shape of Bilton's St John the Evangelist Church tolls its solitary bell. Stay quiet and still for long enough and in an animal's eyes you become field. Young rabbits bounce out from a hedge and run after each other in play, startling me. Two older does sit on the perimeter barely three metres away, fixing down everything in their motion-sensor vision. *Move and we'll be gone.* I keep my head still but angle my eyes to see they have taken up cooperative positions – one at a right angle to the other so they're covering every cardinal point. Every possible direction of approach falls into their beams. Neither is concerned when a line of red-legged partridge slides single-file past their noses to feed.

Ahead, where the wood extends along the river, scratched lines of silver birch and the reddish-brown puffs of trees shake loose their rooks onto the fields, making space for perfect new leaves. Briefly, I'm sure I see a human face in the shrubby ground foliage, a waist-height tangle of messy hair and earthy face staring out over the fields towards me. Somebody crouching. I feel the quickening thud of my heartbeat, blink and raise my binoculars, but it has already slunk back into the shrubs. The morning assumes a brown-yellow light, like opening your eyes underwater in the North Sea, and the rooks swarm upwards over the trees and away in clouds, bouncing, separating, fretting each other, beak to tail feather. Their calls are the harsh, chippy, shocked sound of a wooden peg being malleted into a wooden beam. When I look back again the rabbits have gone. A stoat running along the field edge searching for kits has worried the warren. It comes within a metre, its slick fur the colour of old engine oil. I lie down, eye-level with it, and imagine slipping with it between burrows, molehills and stems.

Sun and sky appear high, removed, as though behind a pane of glass. The hares grow out of the soil to nibble the wheat and mate along the limits of the field. I watch them long enough to lose all sense of time, and then make my mind up to get nearer. Sliding onto all fours I crawl along one of the margins between the crops. Immediately their alert, doggish faces turn and fix me. Twitch, twitch, twitch...then they disperse and go to ground. I cover ten, twenty, thirty metres. Soon I'm closer than I've ever been to a hare. Whispering distance. Hunter-close. The Jack is facing away from me, invisible but for the finest edge of fur amid the green; I'm almost upon him when I cross the invisible tolerance line and he springs and runs, swerving and banking but still moving fast enough to change shape.

> *Then when you have said all this,*
> *You might go out*
> *East, West, North and South,*
> *Wherever a man will –*
> *A man with any skill.*
> *And say good day to you, Sir Hare.*

II

I'll bet you've never seen a hare in Caffè Nero, have you? It's not what you expect when calling in for a morning coffee. Here I am, though, John Joseph Longthorne, born right here in Harrogate in 1945, sitting in the corner and taking an hour and a half over my large Americano sweetened with four sugars. People always stare. *What's he doing in here?* It's the same routine: a quick look, a quizzical frown, and then the embarrassed turn back to the low-fat blueberry muffins. All over in a second. I'm used to it. To be honest, it's not my face that draws attention so much nowadays, the beard hides my lip pretty good; it's more the state of my clothes. They worry I smell. The girls that work here are never too far behind me with the deodorising room spray, but I wash every day; my shirt, jumper and trousers get a good going-over once a week in fair weather, rubbed up with soapwort – *Saponaria officinalis* – and sun-dried in bushes. All right, I can never get them that bleached-bone clean my mother used to – some things you can't wash away – but it's better than the cancer chemicals people lather all over themselves these days. And I mean the men too. Vile narcissism. A sign of the times, I suppose.

Look at these two coming in covered in the gloop, these ladies backing in with pushchairs. They make a beeline for a nearby table before they see me, but then regret the decision. Not much choice today, though. A silent conference via tight-lipped expressions, the raised eyebrows and reluctant agreement. I smile and blow into my coffee. I could say something, but I don't. No point making them feel bad.

'Love the hair, Lu, really nice,' says one as they park

bums and set down extra-large caramel lattes. 'You pleased with it?'

'Be better if this weather wasn't so shit. It's messed it all up.' Lu jams a bottle into her baby's mouth as she talks, dragging her other hand through her fringe. The other unwraps hers from a muffler. It wakes and cries, all pink cheeks and wide, blue eyes.

'I know. I thought spring were here when I woke. Sun shining, blue sky. Tanning time, I thought. Then this. Pissing down.' Another bottle appears from a bag straight into a mewing mouth. 'Here y'ar, sweetie.'

'Spring, my arse.'

'But that's it, Lu, that's what I mean. Good as it gets. S'why I hate England. Bring on Australia. Three weeks and I'm out of here...'

Their voices are lost in the scented, cloudy, flouncing-in of a wealthy woman, all Clarins, camelhair coat and entitlement. I know her. Hers are the tips in the coffee-bean-filled cup by the till. Pound a day. Minimum. Always dropped in a stack of twenty-pence pieces. Probably keeps them in a change jar like some folks keep coppers. She throws me money sometimes too when I'm warming up in a doorway, but only if she's shopping with her friends. She shakes her head as she rounds the old couple asking what 'pesto' is by a line of plastic-wrapped sandwiches.

'Usual,' she says to the barista tending the almond croissants, sliding five twenty-pence pieces into the cup. 'It's just a quick pick-me-up. Heading home soon...'

Maria, the smiley assistant from Andalusia, tries a reply in garbled Spanglish. The woman's eyes linger on her lip piercings as she listens.

'Hay hom, ju say? No wor today, huh?'

'Mmmm?' says the woman.

Maria has another go. 'Ju hab no yorb? No yorb? Ju lucky, *si*?' Maria says *yorb* when she means job. She always asks me the same question.

'Oh job.' The woman cottons on eventually. 'No. I've too much on running the house. Shopping done for the week now, though, so . . .'

A lull in the muzak leaves her words hanging in space. We all hear them. They can't be recalled as they form into a vision of lonely suburban afternoons. Annoyed, she bites her lip and fidgets with her loyalty card as her skinny cappuccino explodes into creation, then hurries back to the silver Range Rover double-parked outside. Dragging in her coat, she slams the door. Back to the safe, self-imposed schedule of the affluent: gym, shopping, a quick whizz around Waitrose. Pick up the kids. Day done. Week done. Life done. Book a holiday somewhere on the iPad. Somewhere in the Maldives or the Caribbean. Places that will be underwater by the time her grandkids are drawing their pensions. And yet here everyone is, pretending, ignoring, avoiding, fiddling away while Rome burns. Emperor Nero, Caffè Nero – same difference.

Me? I only come in here for the coffee and to shake out the rain. Why else would I? This unholy sterility of the mass-produced, of decal-dressed glass, ranks of baked goods and leatherette seats. This 'I'm-sorry-but-yes-you-need-keys-to-the-toilets' sort of place. Privatised pissing. Steam-filled nothingness. It's all that's wrong with the world. They do *really* good coffee, though, and, like everyone else, I'm a sucker for it. A hangover from the bad old days. Time was when I used to insist the girls in my office got me a fresh cup every hour, on the hour, and I admit it still puts a spring in my step. I fairly bounce out of here, high on the Dominican bean. No matter how many times

I've tried brewing up blanched and ground acorns over the campfire, I still yearn for the Java.

The New York jazz has kicked in again. Second time in an hour. Sounds you don't have to listen to. Playlists made up from car advert jingles. Non-dangerous opiate music, more dangerous than you can imagine. Men in square architect specs breeze in on the scent of citrus with fold-up brollies and order strong lattes or 'flat whites' – whatever they are. Each checks his mobile phone for emails, tapping his foot to the bleats and blares of a tenor sax. An aberration inside this sandstone Victorian building, but it fits the reverse-engineered exposed brick, I suppose. Hanging from the ceiling via wires are the faux patina-flecked sepia photographs of street scenes that this coffee shop's very existence has abolished. It wears the images of its victims like a cannibal carries the ears of previous lunches around his neck.

Now that makes me laugh. Sometimes out loud, but I can't help it. See, it's all here in front of our eyes, but no one wants to see it. No one wants to face it. No one wants to deal with the come-down. I'm talking the rise of clone towns, the uniform high streets, the dominance of global brands and corporations. I've seen it spread across this island on my rambles. The enslavement to growth. Profit rules. The high capitalist logic. The result? Oh, you know the result because it's on the news every night, piped into your homes while you flick for the next dose of mindless cooking drivel to switch off to. Climate catastrophe, global warming, species extinction, floods and droughts, human displacement, wars over resources. The Great Ice. The Endgame of the Earth. Mull *that* over as you sip your cappuccinos.

What? You think this isn't our fault? You sit there with your coffees thinking, What've I got to do with global annihilation? Tell me this then: how does that coffee get here? How does it get farmed and produced? What resources go into the heating and lighting of this place? Oil, gas, coal. Fossil fuels. Carbon. You're complicit. The lifeblood to the functioning of this whole mad circus is the very thing causing our planet's death. And who's on Mother Earth's side? Who's policing this wilful destruction? Governments? Brands? Businesses? Telling *us* to recycle and turn off the tap when we're brushing our peggies? Don't make me laugh. Smoke and mirrors. Seven of the ten largest corporations in the world are oil and automobile companies. That's right. They control the means and the distribution. Expand or die. Drill the Arctic. Frack the seabed before rival corporations or countries get there first. Carbon junkies, the lot of them, all high on their own supply and keeping us fat and quiet with jazz, coffee and Range Rover dreams. Or worse, controlling us with depravation, poverty and the vainglory of lottery wins and talent shows.

Truth is that too few care about what happens in the future or what went on in the past. Yeah, I'm talking about you lot, you coffee-shop clichés. I'll bet not one of you knows that this place was a dance hall once and that over there by the sugars doomed soldiers lit cigarettes for hotel girls. Only I remember when the strip club at the end of the street was a Salvation Army Citadel. One day the seas will rise up and this'll be a coastal town with nothing but ocean to the east. Oh, yes sir, I'm mad all right. Mad as a March hare. You should be too.

It's OK, Maria, don't trouble yourself; I'm leaving. Rain's stopping anyway. By the time Miles Davis starts on that horn again I'll have drained the last drops of this Americano and leaped away from your Sani-Talk, your branded cups, your free Wi-Fi and your feedback cards.

Down the high street I'll catch a glimpse of myself in the mirrored windows of that international fashion store. Brown, thin, long-legged, wild-eyed, face dusty as a moth's wing, my hair matted into a bird's nest. My gaze will default to its usual position, my lip, hidden under the furze of a grey muzzle. All the time the blackbird in that ornamental cherry outside Greggs will sing. Paths to freedom, songs to the gods lighting up the sky. Oh and you girls – yes, you two mummies with your caramel lattes – you were right about one thing: spring is here.

Thanks to me.

I was the youngest of four brothers but the only one blessed with cheiloschisis. People tell me that the preferred term nowadays is 'cleft lip', but back then even facial surgeons called it a harelip. In the end it doesn't matter how you dress it up, it's the same thing: a fissure, a rift in the tissue of my *labium superius oris* that happened before I was even born, a non-union that occurred in the womb. I often wondered why it was *hare*-lip. Turns out it's because, as they get older, the skin that covers a hare's gnashers splits into two, exposing their front teeth. Oh, they had a fair go at stitching mine together, the doctors – best they could at the time – but it's always looked funny. The scar gives me a sneer when I smile. Mother hated it. *Poor John*, she'd say. There was superstition in her Irish blood; too many folk tales floating around somewhere back there about hares and witches. She tried to hide it, but I was always aware of something behind those flyaway specs. Disappointment probably. Father was different. He made jokes, but that was just his way with everything. He loved me every bit the same as my brothers; in fact, he insisted on my education first and foremost, even above theirs. He used to read to me every night, sometimes in the mornings before school if he'd finished a late shift on the railway. From as early as I can remember he celebrated this old lip. Called me 'Sir Hare' after an ancient poem his grandfather and he used to recite before they'd go out poaching. He bowed to me every time he said it. 'Sir Hare, *an honour*,' he'd say. And the name stuck.

Maybe it was because of this lip, but I worked hard, a damn sight harder than my classmates at Bilton Endowed School. It's a nice detached now, a smart one too, but it's heartening to see the old waffle-glass windows still run with condensation on cold mornings. And those gnats from the sewage works besieging the laurels like returning holiday-makers. While the other kids galloped around in the grass meadows out back, I used to stay in and practise my speech, writing and reading. And it paid off. When the school was built in 1793, the only condition of entry was being able to read the New Testament. By ten I had memorised all the gospels and could recite them standing on a chair. 'Look out,' Mr Wallbank used to say. 'Here's Sir Hare, the Bible Basher.' But he was proud of me. He even gave me a book token for stepping in and recalling the whole of Luke 12: 16 to 31 at Harvest Festival after Celia Taylor did her ankle on the transept steps.

Growing up I was hardly spoiled for friends, but I didn't care. After school my brothers and I had the run of Bilton, all the way through willow wood, over the railway – still puffing with the narrow-gauge coal engine in those days – and through the meadows. We'd hunt for adders in the hay fields and go bird's nesting. Best was finding a lapwing's nest down in the grass. There were plenty back then in the fields, little straw bowls with their black and yellow speckled hoards inside. Even in those days I'd insist we put all the eggs back. Used to drive my brothers mad. Arms gashed and our lips purple from the blackberry bushes, we'd pick eyebright, ragged robin, crane's bill, speedwell, poppies and cornflowers and, on hot days, swim off the sweat of our labours in the cool, brown waters of the Nidd.

Of all us Longthorne boys I was the only one likely to go to university and I did, on my father's insistence, in 1964. By then Mother was a wreck of barbiturate- and television-dependency only a few months away from being lowered into St John's churchyard. Dad had retired and was slowly becoming indiscernible from the herbaceous border of his beloved garden on Red Cat Hill. That's where we'd sit together and talk flowers, plants, insects and birds. One afternoon when I was reading in my bedroom, he called out to me and lifted up a flint arrowhead unearthed by the edge of his trowel. I told him that, given the short supply of flint in the region, it was most likely brought from the south or the east and used as a stabbing spear for fishing. They wouldn't have thrown flint away. Too valuable. He put both his arms around me. 'You run, Sir Hare,' he said as greenfinches flocked the bird table over his shoulder. 'Go to university. See the world a bit. It'll do you good.'

London was already getting drunk on the good times when I arrived, although I didn't realise it at first. The early months were still a black-and-white 1950s world for me, a miasma of flannel ties and woollen suits, received pronunciation, smart hair, sensible careers, dinner-dances on the King's Road; if you knuckled down, there were promises of one day making the down payment on a Mark 2 Wolseley Hornet and a detached in Surrey. My digs were on Gower Street, a hop, skip and a jump from UCL and the hallowed corridors of the history department. If pushed, I could roll out of bed and be in lectures in ten minutes. And, as that decade erupted into glorious Technicolor, I regularly tested those timings, often with a swirling wine headache and the taste of Avon's *Unforgettable* still fresh on my tongue. My record was stubbing out a post-coital cigarette to presenting a paper on Saxon polytheism in less than eight minutes. I've always been a fast mover. Believe me, my successes with the fairer sex were as much a surprise to me as to my close-knit group of bookish pals, whom I'd tease with tales of removing bras one-handed in the library. I'm sure that most of the time the girls liked me because I listened. The lip worked in my favour too. Damaged goods. They liked all that in those days. It was a good time to be different, and own a stack of early Lead Belly records. My nickname was reborn; I was known as 'Sir Hare' on campus, this time for entirely positive reasons.

If you were lucky enough to graduate in the summer of 1967 you emerged into a new land that fizzed with a kind of limitless freedom and opportunity. England's fields were virginal. Ours for the taking. At least, that's what the Dean told us. My exam grades ensured no shortage of job offers either. Even notoriety gleaned from nights holding court in Soho's music clubs couldn't dent the reputation. In fact, if anything, it helped. Sir Hare was a 'solid chap' with the intelligence and constitution to out-drink Eric Clapton at the UFO and then, still spinning with the final flourishes of lysergic diethylamide, ace an exam the following after-noon on the consular functions of ninth-century witenagemots. I was courted left, right and centre – the civil service, law, even the Foreign and Commonwealth Office sent their suits knocking – but I rejected them all out of hand. Before leaving academia I'd made up my mind to immerse myself deeper into the glorious Dionysian madness unfolding all around me. Those were good days. Days of innocence. The happenings, openings, music, parties, drugs and wonderfully experimental girls erupted together in a throbbing, seething nest of counterculture and creativity that spanned a radius of four miles around Oxford Street. My means ran out eventually, but if you had the brain, skills and charm, business seemed an easy game to master. Anyone could see there was money to be made. The communication revolution was hitting full swing and suburban dreams were engorged by the wealth and convenience of limitless oil reserves from pliant foreign allies. Cheap, centralised electricity was slung across the country. Our unstoppable consumer culture was swelling faster than the liver of a foie-gras goose, fed by TV, radio, newspapers and magazines. It occurred to me that the world of advertising had 'Get Rich Quick' written all over it. *Tap into that, Sir Hare*, I thought to myself, *and you have it made. Easy Street.*

The agency I founded, *SHA!* (I never divulged the 'SH', but the 'A' was for 'advertising'), was an immediate success. By spring 1970 I was signing an office lease covering four floors in Charlotte Street paid for by a client list of thirty-plus major corporations this side of the pond. A fair few the other side too. Things get hazy around the same time we ran the ad campaign for the Rolling Stones' *Sticky Fingers*. Everything began speeding up. I felt like I could move between years as easily as opening the doors of a corridor. Heroin-fuelled months would disappear in minutes. I have blurry recollections of delivering lengthy monologues taken from the *Anglo-Saxon Chronicle* to employees while standing on our picture desk. Someone complained when I devised a series of outrageous perfume storyboards involving ritualistic slaughter. Then there were the obligatory post-work private members' club lock-ins. I let the reins slip, but always managed to hold on just enough. Meanwhile, the drugs got dirtier, the sex more aggressive, the deaths more frequent. By the mid-seventies most contemporaries had moved out to follow a new rural dream of radishes and rotavators. It was *The Good Life* and all that. Me, I continued to chase a particularly destructive narcotic habit around Fitzrovia for God knows how long. My father died at some point but I have no memory of attending the funeral, although I'm told I did. That must have been 1978 or '79.

Oh, I knew the game was up but who has the willpower to walk away from that? Certainly not me. At least, not then. It was like one long scene from *Caligula*. The epiphany came when driving home with a girl around dawn from a party in Weybridge. Sally or Sandy, I think her name was. No, it was Sandy. Mortimer. Nice girl. Married an MP in the end, I read later in the papers. We were both wired on Courvoisier and cocaine and I was gunning my beloved Lotus Esprit around the back roads somewhere near Meadowlands on the A317. The headlights picked out a shape frozen in the road. It was just there for a second – then thump. We skidded to a halt and after a moment Sandy started screaming. I remember the thick grey mist as I threw open the door and staggered back. I remember the smell of the exhaust. Lennon's '(Just Like) Starting Over' blaring out from the stereo. And there he was, lying in the road.

'What is it?' Sandy yelled. 'What is it?'

'A hare.'

'Oh, thank God,' she said.

Thank God? How could she say that? I could feel tears coming; my chest was tightening. I sank to my knees and tried to stop shaking. You see, I saw myself lying there in that road. It was *me*. It was my flank heaving like the clappers; it was my legs all broken. Blood was bubbling from its nose, over its lip. And that eye. I saw the whole world in that eye, like a mirror. I understood everything in an instant. I saw what I'd become but even worse than that, I saw what we'd become. I was ashamed. I picked up the hare and walked off into the fields, carrying it in front of me like a sacrifice. Sandy was going crazy at me – *What the hell are you doing?* I shouted to her to drive herself home. Told her to keep the car. Keep everything. I had no need for it any more. I think it was somewhere near Staines when I realised the hare had stopped wheezing and I buried it in the soil of a rapeseed field and waited for sunrise. In the glow of a new morning I headed north, subsisting on the little cash left in my wallet until I threw it into the River Meden just north of Markham Moor.

I'm not ashamed to say that I cried frequently on my slow pilgrimage through this country; I cried for the way-wardness of humankind and the part I had played in it. Huddled beneath an old canal bridge I watched links of multicoloured traffic splashing endlessly over the rain-washed asphalt. I slept under the murky concrete of flyovers and crossed septic streams that reeked of disgorged chemicals. At dusk, I saw how the houses and tower blocks shimmered with wasted energy. I passed power stations and factories whose spewing funnels arrogantly hazed the sky with carbon dioxide, methane and chlorofluorocarbons. Rummaging for sustenance among field edges, I found the dock, plantain and bindweed dusted white and dying with pesticide residue. The corpses of harvest mice, titmice and sparrows lay limed in mass graves under hedges. Mixed in among them were crushed Coca-Cola cans, the sodden and torn bare-breasted pages of the *News of the World* and the flayed black skins of burst tyres. I was walking through a grey, choking world of the infantile and facile whose caretakers had become sick with greed and lust. I saw a paved-over land populated by people whose minds had forgotten the very ground they came from, who took the gift of life for granted. Humankind was atomised, enslaved, forced to compete with one another for jobs, for housing, for education. Numbed by alcohol, drugs, golf club memberships and shiny falsities, there was no thought for our place as part of the greater organism. Mother Earth had reached breaking point. Even she could no longer stand idly by and watch her spoilt son tearing apart the biosphere. I saw her rage manifested in the eyes of a poisoned sparrowhawk flapping out its death throes on the verge of Common Lane, a mile to the east of Tickhill. *My love is not unequivocal*, I heard her scream in the clap of thunder that broke overhead. With each footstep my vision and understanding became clearer until, around Doncaster, sleeping rough in a belt of larch by the A1, I woke with everything straightened out in my mind. That was Saturday 21 March 1987.

By the time I tramped back here to Harrogate I was physically exhausted but my mind was gripped with righteous energy. I set about explaining the need for a new global ecological consciousness to the bored commuters parking their Rovers and Volvos at the railway station. Climbing onto the footbridge's handrail, I hailed them from high over the tracks: 'Brothers and sisters,' I said, 'stop what you are doing before it's too late. Everything in our sick civilisation is either made by or moved by fossil fuels. Fertilisers, pesticides, building materials, medicines; your fine suits and clothes, your power, your heat, your light, these transportation infrastructures, all are made of carbon. This will be the world's death. It will be your death...'

I didn't get much further before the station staff hauled me down, restrained me and issued a platform ban for life.

I tried more subtle approaches after that: knocking on residents' doors or shadowing shoppers as I weaved through the crowds of the high street. I begged and pleaded that people looked at themselves and the world and recognise what was happening. Some laughed; some shouted; some pushed change into my hands; others punched, kicked or urinated on me as I curled up to sleep among the flat-pack boxes behind Woolworths. There were spells in hospitals and police cells. Once, when I was stricken with flu, an elderly lady kindly let me sleep for a week in her spare room until her son, a local estate agent, found out and threw me into the road. I took to handwriting pamphlets and petitioning the local council. Surely they'd see sense? But when reception staff began to routinely call the police upon my arrival, I adopted more guerrilla tactics, hiding out in the elegant gardens of local Conservative MPs' houses on the Duchy Estate and flyering their cars each morning. I was threatened with arrest on charges of vagrancy. Then trumped-up charges of sexual deviancy. In my absence, steps were taken through the county court to detain me permanently at High Royds mental institution on grounds of lunacy.

My name then, if I had one at all, was 'madman', 'tramp', 'good-for-nothing', 'blight'. I was chased and spat at, but that was nothing compared to the savage beating inflicted by a group of drunk football fans enraged at England's semi-final exit from Euro '96. Disoriented and half-dead, I crawled with my meagre belongings to the only place I knew no one would follow, where I could shelter and recover unmolested: the woods and meadows beyond the new urban limits of Bilton. Here, in the remnants of the realm that had once been my playground, I fell back on the deep-rooted knowledge accrued through my father's

botanical obsession and my own historical studies, tending
my injuries with the herbal remedies of the Romans. I used
sphagnum moss to clean the wounds and then stuffed them
with foul-smelling hedge woundwort – *Stachys sylvatica*
– a legionnaire's cure-all I discovered to be incomparable
in its efficacy. The bleeding stopped and my skin healed
without infection. The fever that followed subsided quickly
and a tea brewed from its purple flowers restored me to level
spirits. It's true that I'll always carry a limp but at least
I was able to walk again. Then, as my strength returned, I
haunted the same spots I had as a boy, surviving on food
I could forage, steal from supermarkets (on point of prin-
ciple) and blag from the North Outfall allotmenters who
tilled the soil between my old school and the sewage works.
Each evening I lit a cooking fire on the bank above the old
weir and watched the flickering shadows of Daubenton's
bats hunting insects over the Nidd.

Readmitted to the societies of the wood and field, I fell
into the rituals of earlier times. I prayed to the gods of
Celt, Saxon, Angle and Jute and communed with the
insects, animals and plants seeking to learn the secrets of
their microbial, cellular majesty. I dabbled with mind-
altering concoctions of liberty cap – *Psilocybe semilanceata*
– and pine needles, stripping away my pain and loneliness.
All those hours of research at UCL proved invaluable in
this homecoming and I dredged my memory for ancient
practices to enact among the trees. The ebb and flow of
birdsong, the rise and fall of the sun, such things became
my world. The slow spinning of the earth, the circadian
rhythms of the solar day, the life and death of the flowers
and the fruits, these whirred the mechanisms of my mended
biological clock. I moved barefoot with the wax and
wane of the moon, synchronising with the lunar month,

realigning my entirety to the rotation of the planet in the universe.

In the middle of the wood is a small clearing that becomes a crucible of sunlight in summer. It is a celestial spot but hidden to all that don't know it, so it was a shock one day to find a man standing there as I returned from washing. Wearing a cardigan and tie, he was picking about my sleeping bag with a stick, oblivious to the glorious light and occasionally looking about himself like he'd lost something. It took a while before I realised it was my eldest brother, George.

'What the hell are you doing down here?' he asked. 'We're all worried. Joy says she saw you shoplifting in Asda last week. For Christ's sake, John, look at the bloody state of you.'

I told him everything, hoping he'd understand. I laid out my thoughts on carbon societies and greenhouse gases, and why the myth of unending growth could only result in destruction. I explained the danger of indifferent, short-term political economies and revealed my insight into the micro-ecosystems of the hedges and woods. I thought he'd been listening, but evidently he'd been growing angrier. Spittle flew from his mouth as he told me he was a prospective councillor for the Harrogate Bilton and Nidd Gorge division in coming local elections. He said my living there could ruin his chances if word got out.

'You better not eff this up for me,' he yelled. 'I've worked too hard for it. I sweated blood in the office while you were gallivanting around London high as a kite.'

The *coup de grâce* was his assertion that my presence there was illegal, that, by law, I could be, *should* be, removed. We argued for a while and I accused him of colluding in the widespread exploitation and destruction of this world. He said I had sustained a serious head injury and that I needed proper care. Round-the-clock care. The sort of care where they lock you away. 'You're ill, John,' he said. 'Time to stop with these daft ideas and speak to someone.' At these words, I asked him to leave, which he did after throwing a crumpled twenty-pound note on the ground and calling me a damn fool. Off he stamped through the trees, swearing. I watched him go and realised that I was speaking those passages from Luke out loud, just as I had on Harvest Festival all those decades before.

The ground of a certain rich man yielded an abundant harvest. He thought to himself, 'What shall I do? I have no place to store my crops.' Then he said, 'This is what I'll do. I will tear down my barns and build bigger ones, and there I will store my surplus grain. And I'll say to myself, "You have plenty of grain laid up for many years. Take life easy; eat, drink and be merry."' But God said to him, 'You fool! This very night your life will be demanded from you. Then who will get what you have prepared for yourself?' Do not worry about your life, what you will eat; or about your body, what you will wear. Life is more than food; the body more than clothes.

The winter that year was a particularly long and bitter one with deathly nights and short sunless days of rain, frost and sleet. It showed no sign of letting up, and by the last days of March I was ready to go further and consign my form entirely to the divinity of the radiant dawn, Eostre, if only her up-springing light would come again to warm us beleaguered survivors. Pledging to do her will and become a living spectacle of her joy and blessing, I built a bonfire at sunrise on Easter Sunday, 2007. As the smoke filled the canopy of the wood, I heard St John's Church ringing out the good news of the resurrection over the town, over my parents' greening bones. I thought of my father sitting in his garden and smiled. It had been a long time since anyone called me Sir Hare, but I felt myself again. I knew who I was. *What* I was.

That evening hares flocked in the fields. A drove of them, maybe thirty or forty, were gathering in the gloaming. I rushed from the trees to join them, tearing over the soil in giant leaps. We ran together east of here along the edge of dense hazel thickets, up sloping fields and through hedges until I could run no more and collapsed panting on my back. It was there that I chanced upon a strange relic lurking in the undergrowth, a dome of gritstone buried by brambles, wood anemone and bluebell leaves. I only noticed it at all because of the marshy ground and stench of rotten eggs. Brushing aside the foliage I revealed a capstone covering a spring, potent with the smell of sulphur. On the stone were carved the initials 'JW' and the numerals '1778', a date that had been chiselled just as deeply into my memory by Mr Wallbank back at Bilton Endowed School. It was the year the Enclosure Acts were passed in Harrogate. The enormity of that discovery dawned on me, sending me reeling backwards into a carpet of greater stitchwort. As I stared at it, the simple stone testament of ownership became the embodiment of dispossession. With such markers they had broken up, divided and enclosed our beautiful world, turning fields, springs, trees and beasts into commodities for mankind to claim, buy, sell and kill for. Global enclosure, exploitation, industrialisation, climate change, I saw all of it radiating out from that capstone, leaching out its poisonous darkness over the land. Eostre had clearly led me there for a purpose and I wasn't about to disappoint her.

Sending up a prayer of strength to *Tiw*, god of war, I set upon the stone, heaving and hauling away until I wrestled it sideways into the foliage. From beneath, water bubbled and gushed through black mud. I bent down and scooped a hollow, then drank thirstily, handful after handful, raising my wet face to the moon and shouting my joy and thanks.

They used to say that water drawn on the Easter Day was holy and healing. They weren't lying. I was at one with everything; my head swam with old voices, the deeper music of the stars and the dreams of ages. I was the swift hare running through Eden, hearing the universal harmony that comes in the union of soul, spirit and earth. The divine. And she was there at dawn the next day, Eostre, standing fair, golden-maned, fully formed, alive, gentle, fragrant as honeysuckle, randy, multi-scented, multi-coloured, many-voiced, bright and bold, her warm breath as pure as a baby's. The soft hair on her skin was the waving grass; her face was the perfect pear blossom blessing the bushes. And it was me that summoned her.

I released the spring.

I move around a lot now following the sun. It's necessary also since an arrest warrant for theft and criminal damage was issued in my absence at Harrogate County Court four years ago. I travel on foot, always alone, up and down this country, from village to village, on the back roads and down the lanes, over the fields, bedding down in the spinneys and copses. You've probably passed me someplace at some time, glimpsed a familiar shape in the hedge, a form in the wheat. I'm there at the Summer Solstice at Stonehenge and Winter Solstice on Windmill Hill, overlooking Glastonbury Tor as the glowing orb rolls along its ridge. Sometimes, when the weather gets too much for these old bones, I'm forced to warm up or dry out in the public libraries where, up against a radiator, I'll read with a mix of hope and horror about climate conferences or the UN's embracing of progressive energy agendas, then inevitably, a few pages on, the issuing of fracking licences to multi-national corporations. On the occasions I can beg the change and bear the hypocrisy of my actions, I'll take a corner seat in a coffee shop and spend an hour and a half over a sugary Americano. Ah, it's heavenly, but nothing compares to the waters of that Bilton spring and the vigour of bringing life and hope back to this dark world. That's why I defy the law, returning here at the same time every year. I *have* to, you see. Nature's order must be revealed. It's our only hope. I prepare myself by watching the run of the hares over the fields and then, when the moon is right, I replay the ritual and release the spring again. After I've drunk my fill, I set up camp down by the weir, rolling out my sleeping bag and sparking a cooking fire, just like I used to. I can sleep then, safe in the knowledge that tomorrow the light will return.

III

I leave him there curled up in the earth by the weir in his grubby brown sleeping bag, a union of land and landless. I scramble up the wood to emerge fairly soil-covered myself into a shock of daylight. While my back was turned, the blackthorn has blossomed. Its flowers are heaped over at the meadow's hedges like snow, bright enough to burn your retinas. They're a Tate & Lyle white, the white of wedding dresses and meringues, bleached against the contrast of their black, leafless branches. A crow is suspended for a second in the sky above, held as if on wires. From that height the sloe flowers in rows must resemble chalk on a football pitch, parcelling up the fields, splicing the meadows, woods, road verges and houses. Segmenting them further are the shadows, the shifting lines of the high sun. More divisions, more borders. Today all these lines make me feel sombre and shut out. I'm sure it's *his* influence, Old John snoring away down in the trees, but perhaps all his talk resonates because he has a point. I trudge around the fields, keeping a lookout for farmers and hares, but seeing neither. Sloshing the clay off my boots in Bilton Beck releases dun-brown clouds of liquid earth, washing away downstream. My thoughts turn to the land and its ever-changing states. The truth is that a spectre hovers over all England, a ghost from a time when people were rinsed from fields as easily as mud from a heel. I've been aware of it since I first came to the edge-land, ignored it perhaps, but it still casts its shadow – over the town, the country and this space between. Time I searched out its relics; time I traced its source and looked full into its face. And now I know where to start.

*

Finding the spring proves harder than I thought. There is little mention of it in the definitive Victorian study of the area, William Grainge's *History and Topography of Harrogate and the Forest of Knaresborough*, and it's a similar story in more recent regional histories – a predominant concern with the town's famous wells, the springs that sprung the spa resort. I find tantalisingly vague references claiming the purest natural waters are to be found at the outlying 'Bilton spring' or, more grandly, a 'Bilton Well', but no indication of its position aside from references to it being on private land and (with the verbal equivalent of a waft of the hand) an assertion that it lies 'somewhere near Bilton Hall'. As geo-location goes, it's not exactly pinpoint.

So I approach Bilton Hall on foot down a gravelled driveway edged with daffodils and I stumble across my first butterflies of the year. Two tortoiseshells flit between the edges of the docks and dandelion flowers on the lawns. Above me, rows of trees touch twigs to create a thin canopy filled with singing finches. A passing car kicks up the chippings; the driver nods and raises a finger in greeting. He is wearing the recognisable tunic uniform of a medical professional; the hall is now a private nursing home and an expensive one at that, the sort where you might expect doctors mustering at the press of a button. Its landscaped grounds and handsome Jacobean-style brick and stone exterior are well preserved, beautiful even, retaining that dreamy, mournful tranquillity you find with old manor houses. Enclosed by empty fields to the south and east, and farms, woods, river and the edge-land to the west and north, it has the air of luxurious seclusion. Less than half a mile from its door are the shuttling A-roads and petrol stations, rubbish dumps and back gardens, the houses, churches, hospital buildings, garages, salons, supermarkets, bathroom stores and autoparts retailers that now conjoin eastern Harrogate, Starbeck and Knaresborough into one continuous colony. Yet from here all that feels like a war being fought in foreign climes. The hall has a different, slower speed, like a record on the wrong setting. As you approach you're enveloped by a sense of isolated history, an accumulated repository of dormant and dying memories being relived over and over again. It's there in the expression of the residents who, at just gone 10 a.m., are now caught between breakfast and morning tea and sit dozing in wingbacks or stare out the mullioned windows at the flowering magnolia and rhododendron. It's there too in the fabric of a building sited in this same position in one form or another for nearly seven centuries.

'This is my blank face,' says Elaine, the sister on reception, as a loud buzzer parps somewhere down a corridor. 'I've never heard of a spring here. You mean a water spring, right?' Another parp. 'Becky?' she shouts over her shoulder. 'Springs?'

Becky pokes her head around from the back office chewing a pen, sticks out her bottom lip and shakes her head. 'Not that I know of. I'm pretty sure this whole area used to be a deer park, though, back before this place was even built. There's a map on the wall over there might help.'

Round the corner in a tatty frame hangs the faded, sepia plan of the hall taken from an old set of particulars. It is of little use, but Becky was right about one thing: go back far enough and hall, grounds, woods, river, all of this was part of something bigger, the huge 20,000 'Forest Acres' (roughly equivalent to 30,000 acres as we now define them) of landscape that constituted the Royal Forest of Knaresborough. The precise date of this enormous sovereign land grab is lost, but it was probably around the time of William the Conqueror's parcelling out of English lands among his followers. By the end of Henry I's reign in 1135, the forest was regarded as a highly valuable asset with boundaries that extended from close to Knaresborough Castle westward over 160 square miles. The word 'Forest' can be misleading, for although cloaked in dense woods, in places it wasn't solely tree cover, rather a mix of woodland, clearings, rivers, open moors and heath given over to the proliferation and protection of 'venison', a broad term that included all game – wild boar, wolves, grouse and hares. Primarily, though, it was red and roe deer that served as the 'noble' quarry for the Norman and, later, Plantagenet bloodlust. In the Forest of Knaresborough two specific areas were fenced off especially for their concentration using a system of sharpened palings that allowed game entry but limited exit. These corrals or 'parks' were therefore prime spots for the hunt, carefully maintained and policed by 'Foresters', the local workers who lived on and around the land. These men enforced William's brutal forest law over the rest of the community, which, as the *Anglo Saxon Chronicle* tells us, 'set up great protection for deer and legislated to that intent, that whosoever should slay hart or hind should be blinded'. One of these condensing areas, Haverah Park, lay to the west of the village of Beckwithshaw, three miles south of what is now modern Harrogate. The other covered the verdant woods south and west of the River Nidd, including all of what is now the edge-land. Its heart was the ground I'm standing on. This was Bilton Park.

In 1380 John of Gaunt, son of Edward III and then Lord of Knaresborough, ordered the construction of a new hunting lodge. It was this building that mutated and morphed over centuries into Bilton Hall. Despite the thrill of the hunt falling out of favour among successive royal dynasties, The Crown retained possession until it was sold by Charles I in 1628 to help finance his disastrous foreign policies and ongoing war with the French. But even prior to its sale, the noble and wealthy Slingsby family had been leasing this estate. Theirs was a name that would become synonymous with springs around Harrogate following Sir William Slingsby's 'discovery' of the town's first mineral water source. Returning from a Grand Tour of Europe, he realised that the iron-rich water around the Tewit Well, located in the marshy meadows a few miles away to the south, had the same properties as those he'd enjoyed in the Belgian town of Spa. It was a revelation that would birth not just Harrogate's reputation for restorative waters but also the word 'spa' in the English language. (At least, that's the story Harrogate's local historians stick to; similar claims can be found in most spa towns in Britain.) Whether Slingsby knew of the existence of an even more efficacious sulphur well in the woods by his home is unrecorded. In any case, his family's days as residents at the hall were numbered. In 1615 charges were levied against his son, Henry, then keeper and 'herbager' of Bilton Park, for allowing it to fall into a dilapidated state. He was accused of felling trees and killing deer, aristocratic misdemeanours serious enough to warrant a heavy fine and the family's swift removal. Ironically, three years after Charles I sold the estate, it was bought by Thomas Stockdale, a staunch parliamentarian and fiercely bitter political rival of Henry Slingsby. Bitter enough that he perhaps even bought it to spite him. In the tradition of primogeniture, ownership passed down the Stockdale line until an heir, another Thomas, mortgaged the hall and land in 1720, investing the

thousand pounds raised in the ill-fated South Sea Company.
When that bubble burst he lost everything and despite desperate remortgages and loans, the Stockdales were ruined
too. They would suffer the same fate as the Slingsbys before
them: the family was unceremoniously evicted and, searching for some kind of renewal, emigrated to the New World.

A cacophony of buzzers sound, snapping me from my
thoughts. Urgent calls for attention from residents' bedrooms upstairs. Harsh, robotic yells, like ship alarms;
reminders that enforced removals to new worlds are ongoing
at Bilton Hall. Rushing off to answer them, Becky and Elaine
suggest I talk to the maintenance staff outside. 'If there's a
spring anywhere around here, that lot'll probably know
about it,' says Becky, throwing me a wave. 'Happy hunting.'

Alone, I loop around the manicured hedges and wooden
sun-recliners. Visions encroach on the neat lawns and apple
trees of tweed-suited hunting parties meeting for morning
sharpeners, long pre-war summers where the white-dressed
daughters of the hall pick flowers to brighten the cut-glass
table decorations. Behind a plastic shed, I find a group of
Polish groundsmen sweating as they load wheelbarrows
with turf and compost. I'm met with more blank faces at
the mention of a spring, but when the boss arrives he shows
half an interest. With a shrug, he grants me the freedom to
roam. 'Knock yourself out,' he says in a thick Scottish
accent. 'But let me know if you find anything. I could do
with making my fortune.'

Stepping over flowerbeds blooming with primroses and
down through the overgrown margins, I cross into a line of
holly bushes and oaks. From my jacket I pull an illegal
photocopy of an old map rooted out from the bottom of a
drawer in Harrogate Library. It marks a patch of the sloping
woodland north of Bilton Hall's grounds and leading all

the way down to the Nidd as 'Spring Wood'. It's a stab in the dark, but it's the best I have. Blundering on through the scrub, leaf litter and rhododendron, I stray further into trees that have clearly been left to their own devices for decades, maybe centuries. Private and unused, the wood is a deserted place strewn with lines of collapsed fences and the sudden shrieks of pheasants. The air seems to descend slowly, like the ground, sinking away from the hall and the world behind. There are young silver birch and hazel, the progenies interspersed with the rotten hulks of their fallen brethren. Oaks flushed with youth give off the warm aura of innumerable new leaves. Behind, beeches grab at the sky with furry, green limbs, like mould-covered bones. A woodpecker drills away. Soon trying to marry the map's markings with such deep cover becomes futile. I'm surrounded and short-sighted by similarity.

Branching left, I reach a high, unkempt grassy field. I can't tell if I'm still on the estate grounds or have wandered into some other private plot. The boundaries are disorienting; more fragments of low walls, wired-off muddy-brown patches and enclosed clumps of reed and cotton grass. I climb a sturdy wooden fence, this one blond and splintery, recently sawn and nailed together with shiny barbed wire stapled along its top, and come to a fold in the land. The rumbling drone of big machinery roars and clangs down the slope. At its top a digger is tearing down the wall of an old bungalow. Workmen in orange high-vis jackets and hard hats are mixing concrete nearby, leaning on spades, waiting to pour the foundations into a wedge-shaped cleft in the ground. Not one of them cares that I'm here. *Smash*. Another stud wall and steel-framed window crumples and disintegrates. I stroll off west, climbing another fence before cutting back into the trees.

I have all but given up looking when a careless sidestep sinks my boot through the leaf cover and into boggy, smelly soil. A tiny stream trickles downhill. Up close it pulses with the living current of deeper earth, gurgling, whispering. Tracing it back through wild garlic and the shiny spurts of bluebell leaves reveals a shallow dip that has something in its middle, a sandy grey-green shape, something large and lost in the undergrowth. An old capstone. It's been pushed aside, intentionally dislodged, and now slumps lazily in the mud, skew-whiff. My eyes move to the outline of the carved initials and a date on its top, but I know what they say even before I read them.

JW 1778.

It's a fact that never fails to jolt my brain: the idea of private ownership of land as we know it today is only a few hundred years old. *A few hundred years*. It says something about the world we live in that even to absorb that information requires a moment of reflection. Just conceiving of it necessitates a mental dismantling of everything that modern existence is predicated upon. Look from an aeroplane window over Britain's patchwork of farms, fields and woods today and what you'll see is an almost entirely privatised place, three-quarters of which is owned by the wealthiest 1 per cent of the population. Swathes of it are even propped up by the taxpayer in the form of grants and subsidies. Yet, in contrast, as Simon Fairlie – editor of *The Land* – states: 'Most of the rest of us spend half our working lives paying off the debt on a patch of land barely large enough to accommodate a dwelling and a washing line.' Ours is a deeply entrenched culture of exclusive land possession, of privilege and poverty, begun with the

Norman land grabs and legitimised by a 500-year system that spread from these shores and would go on to change the physical and psychological landscape of the world. *Enclosure.*

In the minds of medieval peasants, the idea that a single man or woman might one day have absolute and exclusive rights over an area of ground would have seemed incomprehensible. Although The Crown or a Lord of the Manor ostensibly 'held' much of the land, it was worthless without peasants working it, producing crops and paying duty as tenants. In return, they had rights in common over ground, allowing them to sustain their families, which in time became codified, passing down from generation to generation as customary and legal rights of access. They became traditions and outlived the feudal system, their persistence down to the effectiveness of the open-field system. This was a kind of cooperative arrangement between neighbours and families that formed the bedrock of rural communities and villages. England's clay soils were heavy going and dragging a large wooden plough through them to create ridges and furrows might take as many as four pairs of oxen plodding ahead of a ploughman. Few peasants could afford to buy – or had enough land to support – more than one or two beasts, meaning ploughing was always a joint enterprise. It was 'open-field' because hedges would have impaired the manoeuvrability required from the plough; instead, cleared land was divided into fields, furlongs and strips. It was a rigid structure that didn't allow for innovation, but out of necessity peasants synchronised with each other and the earth around them, sharing their oxen, tools and crop plans across their strips, adhering to rotation systems for wheat, rye, barley, vegetables and oats that grew flecked with the confetti of wild poppies and cornflowers.

Fallow fields were set aside and used to feed livestock, ensuring good manure cover for crops the next year. Beyond the plough strips was common land, the vital shared space of meadow, wood and pasture that provided all the other essential resources peasants needed for survival: access to grazing pastures for a cow, hay meadows, wild foods, bedding and acorns for pigs, wood for fuel and withies for thatching.

It was a humble existence, one that would later get distorted by distance and stoke visions of a pastoral Eden distinctly romanticised by the muck-free pen and paintbrush. In reality, peasants were engaged in a brutal, primitive struggle with nature and destined to live hard, hungry, filthy and unforgiving existences at times. They were vulnerable to crop failure and pestilence, to illness, disease and the vagaries of climate and weather. Death was everywhere, the stalking shadow at the ploughman's heels, lurking in the nick of the blade, but these were people who belonged to the land and wouldn't have been hardened against its beauty. Wildlife abounded with no pesticides or gamekeepers to interfere: kestrels hovered over the crops, sweat-stung eyes saw flashes of skylarks, corncrakes and kites; hares, harvest mice, stoats and weasels sprung from their scythes. They would have felt the subtleties of the wind on their skin, they knew the smell of slow, thick autumns and saw enormous red-sun spring dawns and winter sunsets. It's there in the folk songs and the festivals. They knew of equinoxes and solstices and rejoiced in those heady high points in the wheel of the year. They felt a sense of belonging difficult for our removed minds to comprehend or understand. You can be sure of that.

Everything changed with enclosure. In the late medieval years, under the watch of the Tudors, the open-field began to come under attack from wealthy landowners who increasingly evicted tenants to make way for sheep ranges that could provide the wool required for a new and emerging textile industry. Initially, sheep had been seen as a sensible solution to controlling empty and untended land, for which a landowner might receive a fine, but the proliferation of flocks supplying wool to a booming trade opened up a new concept – earth as a profit-making space. Wealth could be won from soil. The rabbit was out of the hat. Pursuing considerable financial interests, the swelling landowning class received support from like-minded reformers and statutes that legitimised claims of ownership. Despite protestations from the Church and denunciations like the oft-quoted words of Thomas More in *Utopia* – 'Your shepe that were wont to be so meke and tame, and so smal eaters, now, as I heare saye, be become so great devowerers and so wylde, that they eate up and swallow down the very men them selfes. They consume, destroye, and devoure whole fields, howses and cities' – rural depopulation and poverty followed on a wide scale. Desperate, the rural poor hit the roads. Communities collapsed and entire villages vanished, fuelling a succession of ill-fated peasant riots and rebellions throughout the 1500s. By the middle of the seventeenth century, half of all agricultural land in England had been claimed and subdivided with catastrophic consequences. Many thousands were dispossessed and the peasantry reduced to selling labour to a rising class of tenant farmers who, requiring far fewer men to manage sheep, no longer needed their work.

If this was the wound, the deathblow fell with the parliamentary 'Inclosure Acts', which reached a peak in the years between 1760 and 1870. Up to that point, enclosures had primarily turned land into sheep pasture, but with the clearances of Scotland for wool, and land grabs in India and the Southern US states for cotton, the advocates of enclosure changed their tune. They methodically redrew the landscape. Reforming open fields, pastures and the 'waste' land into hedged and fenced compact units for arable production, they left only greatly reduced patches like village greens or small commons, turning yet more people off the land. Industrialised England's cities and towns swelled with factories and mills that needed a large and cheap labour force. For the millions of rootless and landless, two paths led away from the fenced-up commons. One was to endure the drudgery and humiliation of becoming a roaming agricultural labourer hired on pitiful and unsustainable wages, dispossessed of rights, slowly being starved by the land you'd been hefted to for generations. The other lay over the smoke-wreathed horizon, through the black hole of the factory gate.

Parliament was a committee of representatives largely drawn from this landowning class, rather than a democratic body engaged with people's rights and it was, unsurprisingly, complicit. As were the ranks of newly wealthy merchants, tenant farmers and neo-elite that rallied to demonise the 'laziness' and slovenly nature of commoners disinclined to work beyond supporting themselves and their families. The word bandied about in the reams of pamphlets and reports produced at the time was 'improvement'. It was also maintained that enclosure was essential to Britain's increasing industrial might and expansive colonial ambitions and, in many senses, it was. This was efficiency over sufficiency, agribusiness over allotmenting. Yet its opponents knew the landscape that lay over the hill – the further engorged coffers of the haves, the have-nots driven into urban slums.

*

Moving my hands back and forth over its weatherworn surface, I trace the details chiselled into the capstone. *JW 1778*. Faded and mute, they can offer no further explanations, no eyewitness insight to the stories here. Little is known either of John Watson, the man who purchased Bilton Hall in 1742, beyond the fact that it was he and his descendants who surgically enhanced the building through its Golden Age, remodelling and improving its ancient skeleton, adding the grandeur expected in eighteenth- and nineteenth-century society: ornate staircases and stone fireplaces, new attics and wine cellars, mock-Tudor brickwork, chandeliers, panelling, façades and bay windows. The capstone tells us that it was Watson's grandson, also John Watson, who was in possession of the hall at the time the lands surrounding Bilton Park's borders went through its own Parliament-sanctioned processes of remodelling and 'improvement'. They are his initials that my fingers run over.

Centuries had passed since Knaresborough Forest had fallen off the priority list of a monarchy more concerned with fighting for survival than the thrill of the chase. Local yeoman farmers had long been granted rights as copyholders and turned the land they cleared of trees over to the open-field and commons system, subletting these privileges to the peasant class. This had been, in turn, further divvied up into strips. Such early incursions of the forest were subject to 'fines', but the fines resolved into regular rents and received the approval of The Crown. Foresters branched out into other industries to supplement their income – flax growing, linen milling, lead and stone quarrying. A form of poor-quality coal, lignite, was being mined in pits in the woods along the banks of the Nidd. Limestone was hewed out and burned to create the slaked lime that could be spread over the fields as fertiliser. Predominantly, though, hands worked ploughs. A map drawn up by the boundary commissioners whose job was to survey and enclose Knaresborough Forest notes the whole of Bilton as tenanted farmland.

We're told that in the grand scheme of things Harrogate got off lightly and, for some, it did. In 1770, the owner of the forest, George III via his title Duchy of Lancaster, petitioned for and was granted the right to demark and fence up the strips and commons, and to auction them off to the highest bidders. After local protests, however, a concession to the town's fledgling mineral spa industry was made. A 200-acre horseshoe of grassland in and around the vicinity of its various iron and sulphur springs was set aside, ensuring the town's hospitality trade could flourish. On paper, 'The Stray', as it is still known, appears a victory of sorts for the commoner. Local stone quarries remained open for a period too; these were rights of access of a fashion. But the devil, as always, is in the detail. Only freeholders or copyholders had legal claim; only those at the top of the pyramids of tenancy received the compensation of land equivalent to their former plough strips. Even then rents increased and enclosure was expensive, incurring legal, surveying, hedging and fencing costs that forced many into selling regardless. Smallholders had legal rights and were likewise entitled to compensation, but the amount of land they were allocated was often so small that, without a commons, it proved useless and also had to be sold. Hardest hit, though, were the cottagers without rights of ownership, but who had survived through informal access to the commons. These people appeared on no records, had no voice to defend them and received nothing when their means of survival was taken.

From the western perimeters of his private park, John Watson must have watched as Bilton was surveyed, divided and sold. He must have seen as cottagers and smallholders were fenced off the fields and families exiled into the overcrowded cities of Leeds and Bradford to become fodder for the industrial age. Some will surely have adapted to the cycles of loom and cog, thrived even; others would have faced the familiar fate of the dispossessed across the world: identity crisis, alcoholism, poverty, death. The capstone doesn't record how Watson felt or – and here's a thought – if the sulphur spring beneath was even on his land. Perhaps fearing the boundary commissioners might claim it, and with no one to contest him, he stamped his mark. Or with visions of a lucrative future spa near his hall, maybe he was seized with entrepreneurial zeal and bought the wood in the land sales that followed. Whatever his motive, he staked a claim and, in the process, froze a moment. Here marks a new departure in our changing relationship with land. For the dispossessed, surviving the shock of subsisting in the city must have required the blanking out of that which they'd lost. Solace can be sought through moving on, through a conscious forgetting; over time land became something other, opposite, for the urban masses. It was the past. We pledged body and soul to new rhythms of production and consumption, slouched over machines, counting down the clock.

The weather turns. A wet wind cuffs me about the ears and stirs the trees. Rain runs off the leaves in its slow *tick-tack* rhythm. I walk eastwards away from the spring and back towards the edge-land, following the Nidd upstream. The journey should be a relatively short one, about a mile as the crow flies, but in the interest of keeping hidden from the row of farms and 1970s bungalows that run along the raised ground to the south-west, I cut through a mess of wire-divided fields, up banks, through hedges and gullies and fenced-off patches of wood. Some places exude such a strong sense of their history that it's impossible not to think about what went before. Every landscape is freighted with stories and their clues, like stray letters or sentences lifted from a novel. And here the story of enclosure runs on ragged and retold through the twenty-first century. Swathes of rectangular grass, flat and uninteresting, are edged with overgrown hawthorns that seethe with rabbits. Stray pairs of shaggy, somnolent horses bow their heads. Soaking-wet trailers and pre-fab cabins bulge on the verge of collapse next to piles of cut logs and the heaped black doughnuts of old tractor tyres. 'Land for Sale' signs stick up from field corners underwritten with improbably archaic agents' names. An off-season caravan park sits closed, waiting for the Easter holidays. And all around is the orange twine of the electric fence ticking away like a metronome, repelling nothing from nothing.

Ahead the field rises into a gentle, tree-less hump from which the land stretches away. To the north is an unending plateau of farmland scattered with isolated clusters of corrugated barns, machinery and farmhouses. To the south, the town's chimneys, antennae and the tessellated slate roofs, turned to black mirrors by the rain. I pull up my hood and press on towards the wild middle ground in-between.

On the nursery wall above the Beatrix Potter rabbit I'm putting up a child's alphabet sticker set. *E is for Earth*, it says. Well if E is for Earth it must stand for enclosure too, for these are inseparable concepts now. In many ways it was this union that kick-started a worldwide process of commodification and dispossession, of debts and dollar signs. Has it all been bad? Far from it. Other letters in the packet spell out the benefits: *H is for House, X is for X-ray*. It's a cold morning, but I can hear the boiler downstairs huffing and puffing. If it didn't, I could call someone to come and fix it. No, it's not all bad, just more complicated.

There's a great phrase. I must've heard it or read it somewhere: 'we gained unimaginable freedoms; we lost unimaginable freedoms'. That's the hand we were dealt, you and I, the game in which our parents and their parents before them found themselves embroiled and, by hook or by crook, staked us in too. What concerns me though, as I peel each letter from its white backing and climb the stepladder, is what kind of world will exist by the time this baby lays eyes upon its grandchildren. *What world will I leave? What kind of game am I dealing it into? Is it any different?* For just as previous generations enclosed the land, the sea, the air, the atmosphere, we're surely enclosing ourselves. Every year we become more insular and inward-focused, at once connected to an amazing virtual global multiplicity yet often detached from the world in any physical, emotional and moral sense. Every decision, from buying food to switching on a light, is mediated by so many logistical, institutional and technological layers that we have no sense of what our actions are responsible for. Our profound alienation from the earth continues. We're the landless and listless, so estranged from our planet, so removed from the decision-making that governs it, so isolated from each other and the life we share this world with that we're seemingly unable even to come together and prevent global human and environmental catastrophe. We're still being divided and conquered by enclosure, only now the fences are invisible and internal too.

Through the bedroom window, birdsong. It is louder than the traffic. Penetrate-your-dreams loud. The sound of spring-morning sunlight, of a world waking. Twittering squeaks and wolf-whistles from starlings massed on chimney stacks and TV aerials; house sparrows fizzing between the little front gardens; a blackbird claiming a shaggy leylandii in the corner of a concreted backyard. I want to run to the edge-land, but Rosie reaches out and holds my arm. It's Easter Sunday. Old rituals to be upheld. Family duties. A roast to be cooked. Chocolate bunnies need relocating from the kitchen cupboard to their overwatch positions on the mantelpiece. Egg hunts have to be conducted over the carpet with my nieces. It's not until sunset, as the family dozes contentedly on our sofa watching *Fantastic Mr Fox* that I can slip away through the door.

Night thickens along Bilton Lane. The warmth of distant blackbird song soothes air heavy with cars returning from family functions. Grey-black cumulonimbus big as mountains run along the horizon. Mountains beyond mountains. A pylon stands stark but beautiful against them, like the mast of a tall ship. There is that rich smell of wood and, from the houses, the clean linen aroma of washing detergent swilling down drains. Our tangle with the land is a messy business. Look close enough, back a few generations, and there's soil under everyone's fingernails. Mine as I write this; yours as you hold these pages. Equally, there are no landscapes in Britain, and few in the world, that aren't managed by our hands in some way. Despite their differences in appearance, rural and urban are not opposing states; our prints are all over both, whether from keeping nature at bay or from nurturing it into being. And there's been good and bad come of these tussles. For the Enclosure Acts heralded a boom for the common hare; it thrived with the control of its natural predators and new fields of more intensively farmed crops to feed on. There are voices today that call for 're-wilding', the removal of the human hand from land altogether, a restoration back to 'natural' habitats where vanished species are reintroduced, trees spring up over sheep-cleared hills and the threatened might thrive. It's a beautiful idea and part of me craves this too; I want to believe this could only be a good thing, but there are nagging doubts: where are we in this picture? Isn't this just another process of clearance? Won't we still have to maintain these places to ensure only the 'right' type of nature abounds? And what about the space required for sustaining ourselves? Where will we grow – and who will grow – all the food our expanding population requires? Mostly, though, I'm troubled by the thought that when acres of wood, fen, mountain marsh, coast or grassland become designated wildlife sanctuaries, they also become another type of no-go area, enclosed by a new class of landowner – countryside managers, conservationists and large charities. No one would question that theirs is noble, well-meaning work, but it's still shutting us out. The human is reduced to the neutral observer, removed and peering in from the prescribed path, the car park, café or picnic table. 'Environmentally friendly' subjugation of land is still subjugation, the process of shaping it to conform to what we want it to be. And who decides at which epoch we stop the clock? Which moment in the story of this constantly changing earth gets preserved in time? Which of nature's myriad forms and ecosystems make our privileged lists? No, it's a messy business this and, whatever we like to think, we are part of the tangle, ever staked at the table. Removing us only redraws the idea of a separating line between human/ nature. And for my money, that's the line that has always done the most damage. That's the fence that needs taking down if we're going to stand any chance of making it.

The sky turns a soft black and prickles with the glitter of stars as I walk from the old railway over the meadow and through the wood to sit at the field's edge again. My feet slither on the sticky earth and I brush along the glowing blackthorn, releasing its musky sweet scent. Stacked up on the hill, the lights of Bilton look like a ferry in dock. From the trees drift the cold calls of owls; above them, like a silver button on a black tunic, sits a freshly polished moon. When they rise from the wheat, the hares look pewter in its light. I wasn't expecting them and don't have my binoculars but can clearly see three limping after each other, feeding then suddenly turning in belts of speed, darting off in one direction and disappearing, only to resurface somewhere new. Moon and hare have long been associated and that's partly because of their inconstancy: just as the hare flits about the field, the moon seems to travel through the sky from one night to the next. The association was also to do with madness. From 'luna' comes 'lunacy'; the hare's irrational behaviour in the mating season is in stark contrast to its usually timid life, causing it to be labelled 'moon-struck' and mad. And yet, despite this, in previous cultures the hare was representative of intuition too, the manifestation of a wise thought leaping suddenly into the mind – the light of knowledge appearing in the darkness. Enlightenment. Madness and genius wrapped up in one.

As with the hare, the spring, the dusk and the equinox, this edge-land is a union of opposites, a meshing of lives human and non-human. It is a paradigm for where past realities meet our now mediated existences. Perhaps there's something in that. Of course we have to live in this modern world, but we must get our boots dirty too. We need to slip between these semi-permeable states, for just as we need our homes and X-rays, we need the intimacy of nature, the empathy and the sense it brings of our participation in a larger, living world. Crucially, we need to be *in* it in order to connect on this physical, emotional level, down among the soil and the stems.

Here on Easter Day less than half a mile from town, hares run through a field of possibilities. Like the land itself, this animal remains enchanting, shifting, imaginary, unknowable, ever-appropriated. Well, let it stand for something else then, a new, modern symbol of balance, personal and global. Find it, watch it, follow it and we begin to experience our own un-enclosure, a vital re-grounding. We come to know the value of every rut, blade and furrow. We may experience our own renewal.

DNA

All this in an eye. I wake. Or rather, am woken. The stirring of limbs in the ashen light as the cold passes from the earth. My pupils dilate in a quarter-second and I lift my nose to the air: death drifts on the bleak wind over the wooded hills, a thousand puncture points of snarling fangs and stabbing spears, as yet invisible but infusing everything, changing the trees and the air, altering scents and colours with the same silent menace of fresh blood blooming in water. The hunt is coming and it is coming for me.

I feel my heart quicken, thump and prepare for flight. Snorting to clear wet nostrils, I stand and breathe, pulling air into my lungs to determine direction, but I find only the clean, safe scents of the forest again. Things are restored. The wave has passed; a cloud across the sun. *Was I mistaken?* Song thrush, wren, robin and blackbird interweave their melodies; nothing seems disturbed. No pigeon flaps in fear from its perch. So I too remain hidden here in the holly and sniff, and sniff. Soon there is the smell of growing warmth as dawn uncurls its long fingers further into the wood, gilding moss and cracked bracken undergrown with fern. An early hoverfly is held in a sunbeam, an insect in amber. I wait as the light rises higher and turns silver-gold, shortening the tree shadows and leaving a trace of itself in the prisms of dew dangling from leaves. A nuthatch loosens a string of them as it streaks along a hazel bough probing for insects, gripping in its beak the spoils, the mess of snapped legs and wings and crushed elytra. Then death touches my tongue again, a charge in the air. *Danger. Closer.* I feel it in the same way a dandelion root senses a fly crawling over its flower.

Concealed, I edge forward to take in the full sweep of my rut. Down from the rise on which this holly grows, through the thick trees and cover, the river is in spate with spring rain. On the large swathe of flatter ground between here and there, grass, gorse and shrub have been thinned and trampled in places by the passage of beasts; I smell bears on these meadows some days. Wolves, too, in mid-winter when they float weakened from the hills to pursue my kind through the river culverts and valleys, bringing down our young, old and lame, tearing open our throats and bellies. It was down there in the meadow in the silver-dark that I walked earlier to feed on hazel leaves, and bilberry shoots; it is there now that a man crouches over my slotted tracks.

He is one with the woods this man, this huntsman; he moves so silently that even the birds don't notice him in his tall brown leather-leggings and thick green jacket. A fleet-of-foot approacher, harbinger, deliverer, tracker, venery scout and shit-smeller; there is more animal than man in his slow-slow actions, bending nose to the ground, inspecting every leaf and stem, rubbing my droppings between thumb and finger. He is a broken-branch-reader brought up as a boy to know the dogs and the forest better than his own kind. At seven years he cut the toenails of the brachets and greyhounds over at Kennelhall and lay down in the

straw among their barking, farting number to sleep, changing their piss-posts and filling their water troughs until he'd
learned the flick of the tail, the bob of the jaw, each of the
hundred hounds' quirks and traits. Apprenticed from page
to varlet to groom, he brushed the horse coats and cut the
dog collars from the stretched dried hide of my kind. He
bathed sore feet in vinegar and walked the dogs in grass-
circles, feeding them bread only to keep them ravenous for
flesh. And those beasts he once served obey him now like a
dreadful god. Even that free-roaming bloodhound beside
him, his lymer, the thick-headed, dead-eyed black and tan,
won't run its rope until the order is given – the quiet *tsssssss*
through the teeth.

The man looks up towards the rise and I know he sees
me, not with his eyes but in the same way that I see him, a
collusion of senses and instinct. He knows I am a strong
roebuck, thick-necked, powerful enough to flee like a young
hart and test his finest hounds. And it is that he seeks for
his masters. The chase. The lymer sniffs and turns its head
in my direction. It craves my lithe perfection. It wants to
master my wildness and tear through my tissues and bones.
I feel a flicker of fear about my chest, but not the sort that
freezes some animals; fear is my kind's gift. It is the force
that drives us and keeps us alert. To lose it is to die in these
woods, leaped on by unseen shapes from the back and the
side, held as you try to run, dragged down into the thrash-
ing maelstrom of teeth and claws.

I stay listening, unmoving, until I become aware that
man and dog are no longer there. I twist my ears and peer
through the brush and high grasses, but they have vanished.
Not a stem shudders in the broken light; the birds murmur
on in circles of notes. A horse whinnies in the distance. It
is as if they were never in the meadow. There is no more or

no less scent than before. I breathe in that essence of a true huntsman: nothingness.

Perhaps this man will be the one to bestow my death wound. Not carelessly with the rage and lust of his pomp-fuelled masters, but with the mercy of one who knows my body better than his own. One who loves my form and will have tenderness in his veins when he kills me. I will kneel moaning, exhausted and gasping and he will approach and push the blade fast and painlessly into my throat, nicking the artery that sends me running forever into the blackness. Then, when my pink tongue lolls through my teeth and my breath ceases to cloud the air, he will let his dogs fly at my neck, briefly restoring in each its former wildness before recalling them for the curée. Another man, a nobler man, a stranger to these woods, will sever my head with a sword and hold it up, pouring my blood over the bread they feed the hounds, then throwing it to the lymer as a trophy. They'll crack my bones under their knees. My entrails will spill over the flowers. They will blow their cow horns in the long notes of death as they tie my hooves around a spear and carry me to their lodge for quartering. The huntsman will take my shoulder for his family and the feast table shall have the rest. Grease will run down chins, like shining steams in the firelight. Yes, perhaps this man will be the one to bestow my death wound.

All this in an eye. Quiet now. Between the warm, soft soil and the closing canopies, the blur of bluebells dazzles the morning. The death choirs of crows sound far away, upriver. Wrens trill in uninterrupted babbles. I remain sharp and aware, but motionless. To move is to be seen, to alert your enemies and become prey. A depth of earth sounds and smells; each I know, each safe. So I lie here and rumi-nate, chewing over regurgitated leaves and bilberry shoots as puddles of light form through the cover to heat my sleek, chestnut coat. Buff-tailed bumblebees haul themselves over the lips of white dead-nettle flowers and everywhere peacock butterflies fall on ramson spears. I smell boar. *Yes, safe now.* A hot, south-west wind rises. *Quiet.* I feel my eyes willing to close and let them.

The bark is faint, but enough, and I instantly grow onto my legs, pulling in lungfuls of air through my nose. But the wind has shifted while I slept and I see them before I smell them. On the far side of the river, the thick coat of larch thins into single trunks as they run along a high ridge of ground, the way fur parts when pushed by a shoulder blade beneath. Moving fast between them are men and dogs, grey shapes bounding and running, then at three short blasts of a horn, they hold their ground. The men squat down among the hounds, stroking their heads, trying to calm them. A pheasant shrieks behind me deep in the wood. I turn as birds burst upwards through the dense cover. Their whistling calls come nearer, passing overhead quickly. Now I hear dogs barking from that direction too, this time more of them, the eager barks and squeaky yawns and yips of pent-up aggression. More horns are blown. I lift a foreleg and twist my ears, but still don't move. *Twitch. Steady.* These dogs won't come, for they are only the relays. The men know I'll bolt with every bit of my strength; they know too that for all their speed, the greyhounds may tire. They're positioning the teams that will take over the chase. *Steady now.* I feel the fear in my limbs, but welcome it and hold it there. *Don't fly, not yet.* Running without sight of them might be to run straight into the pack. First I must know from where they approach.

It is the lymer, the black and tan, that comes galloping through the meadow. I sense the thirst for blood maddening its mind. From where it was seated silently by its master before, it now tears through the scrub, scattering grass as it uncoils its lyam three fathoms and more. It arrows along the same path I trod up here from feeding. Following on a white horse, the huntsman breaks from the trees, his eyes fixed on my holly as he slows to a trot. Taking the twisted horn from around his neck, he puts it to his lips and blows a series of notes. The horse stays; the lymer, reaching the end of its tether, pulls back and snarls, rearing up onto its hind legs. There is barely a moment before other horns sound a response and the trees change into the colourful din of death. In a white seething froth, a plume of grey-hounds streams between trunks and bush the way a river rounds rocks; men in red and blue tunics are close behind, lashing forward their horses, baying, shouting. The hunts-man loosens the lymer from his saddle and throws the leash to a varlet running beside him. With a kick in his mare's flanks, he joins the greyhounds, slapping a yardstick against his boot to keep them true. Sighthounds must lay eyes on their quarry. I know he will be their guide; he will bring them to me.

All this in an eye. A bright eye suddenly more alive than ever. There is nowhere left to hide so I am already running, but after crashing through my holly I have slowed into a noiseless gait, ears forward, clearing the fallen trunks and undergrowth, keeping my stamina. I hear the dogs and horses gaining ground but cannot risk tiring my muscles yet; instead, I bounce forwards in long-bodied leaps, retracting my legs, weaving down a slope to force the hounds through the thicker part of the wood. The dogs must have sighted me, for the huntsman's horse has fallen back and the greyhounds have increased their speed. Paws close in around me, falling like rain on the forest floor.

I let the fear slip into my legs and open my gait to race through the trees, drifting over the banks and bogs as my pursuers dart headlong through the cover, becoming slowed in the mud, brambles and bushes. But they are relentless animals and each time one falls back another shape steals some yards and pushes forward to bite at my heels. We run together like this for long enough that the first wave of dogs slows and stops and a series of horns brings about others. More shouts and the horses are whipped into catching up. My breath escapes in rhythmical snorts; I can feel my heart tiring and my lungs growing heavy, but still a force keeps my legs crossing and uncrossing. My vision narrows into a tunnel, a hazy white circle through which I can better read the landscape and find escape through the thickening trees. My will to live is strong.

Still they come. More dogs. *The relays.* They have crossed the river further up over the wooden bridge. These fresh hounds kindle a deeper fright in me, one I've never known, one I can no longer contain in weakening limbs. I feel terror swell and burst through my flanks, clawing at my chest and squeezing the breath from me. The dogs are close enough that they can sense my fatigue, as though they can see my strength dragging behind me like entrails. One, a black-eyed brute the colour of an old wolf, matches my stride then sprints for my shoulder, flicking its head to snap at my neck, but it misses and falls away. Another is almost among my hind hooves. I cannot suppress the panic that sends me into a final, burning gallop. A small gap opens between my heels and the dogs, but knowing it won't last I bank, changing direction, and plunge through a curtain of thick hazel towards the river.

Led by the foresight of the huntsman, the men have turned their horses to gallop across a clearing and into the same stretch of dark wood. I hear them smashing back through the brush to my side, their steeds' chests breaking saplings and boughs in a storm of muscles and hooves. And always that sound of horns at my back. The hated scent of man. Another of my kind, a doe, springs from its lay and bounds off at a right angle, confusing three of the dogs, but the men scream at them and they swing straight back onto my heels. In a patch of boggy ground between two belts of trees, one of the riders is jostled to the front and he draws his sword and spurs his horse to match my leaping jumps. Twice he hacks down, but I shift sideways so that only the flat of his blade smacks against my backbone.

All this in an eye. An eye bulging with fear, an eye so near to the wide, fast river. But I can run no further. My forelegs are numb and I can't stop them shaking, slowing, giving out. I stumble to my knees, then chest, then fall entirely into the riverine grass and water mint. Ahead the huntsman walks his white horse between the torrent of black water and me. I see the spear resting loosely in his hand and the glint of the gold-edged horn at his neck. *Quiet now.* In my terror, I want to close my eyes and let the flood of fang and spear point wash over me, but my body is already mustering resistance. *Arise.* I leap back up, turn and face the coming horde of horses and dogs, dipping my head, baring my points. My breath heaves in and out in a high whistle. My mouth foams. But I have the will to fight, just as I fought off rivals in all those ruts past, straining and weaving before I ducked and plunged my small, sharp, four-point antlers into their sides, leaving them moaning and kicking out their last alone in the forest. The huntsman shouts at a group of younger men who dismount and tether the greyhounds snarling and barking around me, hauling them beyond the closing circle of horses. The dogs stare with unknown hatred, their long tongues lolling, steam rising from their backs. I read their mindlessness; they are confused.

A dipping sun gleams on the sweaty faces of the riders. They cheer, nod and slap the necks of their mounts. I expect the huntsman to approach with his spear, but it is his master, the one who cut at me, who urges his horse closest and leans over. He wears a fine, sweat-darkened quilt jacket. His face is scratched with thorns, his cheeks smeared with fresh blood. He swings a leg from his steed, steps down, sword in hand, and stands close enough that I notice his fair hair and blue eyes. A beard shaped to his jaw. The men fall silent and raise their horns ready to blow as he steadies the blade at my head; I duck and weave to keep it away, but he keeps its tip hovering below my eye, like a fly. The smells of dung, fear, sweat and exhaustion. Drops of perspiration run down the man's nose. Beside my hoof a wasp sinks its sting into a dying bee. A ladybird stretches its wings on the fronds of a sweet cicely and floats away. I turn my head towards the purer scent of pines and the roar of the fast water. Finally his tongue flashes through his teeth and he lifts back the blade. It is a quick movement, but I move quicker. With a snort, I push at the ground and kick out, summoning the last of my inner fire to bound towards the water, fleeing the roars of surprise and outrage.

All this in an eye. Trembling in my running vision, the huntsman turns his reins and trots to intercept my escape. He becomes even more animal-like as he raises his spear and tilts his head to better plot my weaving, sprinting run. There's fire in him now too. Not malice, but the same intent I've sensed in wolves, an imperative to kill as fast as possible. It's in his tensed limbs, the fluidity of his movement and the guttural sounds he unconsciously gives his steed.

He's ranging and reading, closing in on the large heart beating hard behind my foreleg. With a shout he spurs the horse into a run, but I bank and turn to meet them, bucking and kicking so that we come together in a collision of moving limbs at the water's edge. I am suddenly under his steed's hooves, jabbing my points upwards at the white flailing monster above. It rears and unsteadies the huntsman; he slips so that the puncture-punch of his spear misses my heart and sinks deep into the muscle of my hind leg. An explosion of agony as I leap for the river. Then all pain vanishes as the freezing blackness swallows me. I'm half-aware of another shape crashing into the writhing river beside me, grabbing at my neck and legs. Then it too falls away, like the sky, like the sun, like the land.

No smells now; no birdsong. No noise of men. There is nothing but the thunder of the swollen water as I bob and spin in the power of the current. I feel the long shaft of the spear wobble, tug and tear away downstream and then the warmth of my blood clouding in the cold river. I'm blinded; my lungs crushed as though bitten. Then instinct sparks my legs into a paddle and I break the surface and breathe. Something is almost upon me. Something green. Hands touch my neck. My nostrils flare and I gasp in more air, but then I'm sinking again. When I brush the stones of the bottom with my hooves, I force myself upward, lunging for the far bank.

All this in an eye. The huntsman's eyes rushing up to mine in the violent water, wide and white as full winter moons. He grabs for me again, this time finding my antlers and gripping, dragging us both underwater, down into the churning guts of the river. Swallowed, we corkscrew with the stones and the torn branches, coiled and locked together. I feel the great weight of the man, the heaviness of his jacket and boots holding me down. I kick and buck, twisting my head and, as I do, I glimpse terror. His eyes are like a fawn's when the wolves howl down from the hills. There is no urge to kill me now, only to escape. His will to live is strong. He claws at his clothes and tries to haul his face up to air, but we spin ever deeper. Then he screams; a stream of bubbles before he breathes the river back in, his face screwing up against the pain. Every sinew in his neck strains until, left with an infant's strength, his grip loosens. We separate and as I kick desperately for the surface, he turns with the murky torrent below, vanishing, his mind filling with the faces of his children.

I see all of this in the deep, black, liquid eye of a wild deer. It's suspended over me for less than a second; it will be with me for the rest of my life. I am lying alone in a little hollow under a birch tree twenty metres up the bank from the river when I hear the gunfire. It's mid-afternoon, 3 May, one of those miraculously hot, bright days you get after hard rain. Leaves glow overhead like green clouds, diffusing the sun into friendly warmth. Then from east of the edge-land comes the sharp *crack-crack* of shotguns. I prop myself up on elbows and scan the wood. Maybe it's farmers hunting rabbits or woodpigeons; it doesn't matter, it is too far away to cast a shadow over this little idyll. I slump back into the flowers and foliage, adjusting the rolled-up jumper under my head. Other sounds come and go: the inescapable traffic, a plane passing, a squawking jay from across the river, chaffinches. Then the echoes of four more pairs of shots rumble up the river like rolling thunder. In the lull that follows, there are two new noises, as if someone is crumpling crisp packets in quick succession. The first is

distant enough for my senses to not react but the next –
only two seconds later – is just to my right. Something
alarmingly near, moving alarmingly quickly over the leaves.
Animal. I open my eyes and the roebuck is right there above
me, hanging in mid-air as it leaps the hollow. And in the
same moment I see it, it sees me, for I watch the reaction
twitch along its long flank in a shiver, like a flicked whip.
Time pauses. Before my brain can kick in and muster my
body into a defensive curl, it records details, bits: the lithe,
muscular form as it banks; the black nose and a white chin;
the grunt in its breath; the two small, sharp, twisted, waffle-
cone-coloured antlers and a coat the same hue as oak leaves
in winter. All four hooves are off the ground and its muscle
and sinew are contracting and bulging, rippling and flexing
its chest like sail and rigging at full lick. And then there's
that eye, the surprised, scared, revelatory eye just three or
four feet from mine, looking down at me. I can't remember
ever being so close to a wild animal of this size or feeling
the intensity it brings, as though I've been plucked from an
armchair by a tornado of fur and mass and pungent form.
As though the wood itself is suddenly up and running. It is
something so immediate and exquisite that it's hard to
believe it's happening. And maybe because of this, I fix on
that eye. I see the eye itself, the round orb and glistening
dark surface fringed with lashes, but I see through it too,
past the lens and retina, along the optic nerve and beyond
to somewhere that exceeds understanding; somewhere you
could disappear into and never return.

Time speeds up again, reeling away like line behind a
bitten hook. I presume the deer glides off into trees and
upriver, but I don't see that because my body has flinched
and sought to protect the central, vital organs by turning
itself towards the earth, right arm over my head, legs

scrunched, shoulder braced for an impact that won't come. By the time I sit up again, the wood has returned to how it was. The edge-land is going about its business – the plane still passing overhead, jays; it's only me still caught up in the wake. I feel like a hammer has been swung into my diaphragm. Breathless, giddy, I am somewhere between ecstasy and a heart attack. A heaviness thumps between my ribs and a high note whistles in my ears. I feel twitchy, tense. My body isn't behaving how it should – or perhaps it is. Just as a magnet can magnetise metal when rubbed over it, the deer vaulting me has temporarily animalised my nervous and respiratory systems. There is disbelief, relief, then the urge to shout, to release, vent and share what just happened. I instinctively look around for someone to talk to, someone who – by chance – might have witnessed it. *It was right there!* I want to scream at anyone. *Did you see? It could have killed me! Or* – as it comes to me a second later – *I could have killed it.* And as the human impulses trickle back I suddenly start to feel starkly alone, as though I've drifted too far out to sea while I wasn't concentrating. The adrenalin wearing off. I wish there were other people with me. There *should* be other people here. The next best thing is in my pocket, but my finger hesitates over my phone's screen. From behind me, way back in town, a train flying into a tunnel yawns out a long, discordant honk, the note rising at its end like a disapproving question – *reeeeaaa-lly?* I pocket the handset and feel for my notebook. It's probably right. This doesn't need sharing. Not yet. And not like that. Anyway, what could you say? How do you describe these feelings? How can I explain the kind of primitive excitement flowing through my veins to someone who wasn't here? 'Oh, hi, how are things? I just thought I'd call and say I've glimpsed the immensity of existence.' Because, grand as it sounds, that's how it feels.

A lot of people insist that hunting is in our DNA; some will go so far as to assert it is a fundamental human right, a cultural imperative, even if the necessity for it to provide our food, clothes and shelter has long been delegated to more convenient, efficient or morally palatable methods. Whatever we might think – or not think – of it today, hunting and the process of being hunted ourselves is old programming in the human brain, undoubtedly at the root of the inquisitiveness, admiration, respect, fear and sense of kinship we still feel for wild animals. Prey or predator, the creatures we feasted on and fled from have stalked waking thoughts and nightly dreams throughout our evolution. It's reckoned that between *Homo sapiens* and our ancestors, we've probably hunted animals similar to deer for as long as two million years. The roe deer itself is thought to be unchanged in appearance in the last million. That's a long time to have known one another, a long time to have consigned each other's forms and features to deep memory, especially considering how hands-on and brutal the relationship has been.

Long before its incarnation as a royal hunting forest, the ground I'm sitting on would have run thickly with roe and red deer. Huge numbers of beasts flowed into the Preboreal landscape vacated by the ice as it vanished northwards. Eleven thousand years ago, in temperatures and woodland cover similar to now, Mesolithic settlers – if 'settlers' can be the right word for hunter-gatherers – moved through this gorge, stalking, slaughtering, skinning, cleaning, cooking and eating these deer, sewing their skins into clothes, stretching their hides for tentage, fashioning weapons and decorations from their antlers and bones. Noses knew the smell of eviscerated guts; fingers the feel of fat and the breaking strength of leg ligaments. Such things were routine, day-to-day necessities, the details of our pre-agricultural dependency on these animals. And from the human perspective, it was nothing less than dependency.

You don't need to travel far from here to get an idea how deep the relationship ran. Fifty miles on a rough bearing east-north-east, five miles south from the coast at Scarborough, lies one the most extraordinary Mesolithic sites in the world. The smooth green pasture and plough soil you find at Star Carr today is deceptive: over millennia the build-up of decomposing plant matter remade this landscape, burying under peat all traces of the vast inland lake that filled the area during the post-glacial period. Burying, but not destroying. The peat has preserved much of what is normally lost, like flowers pressed between pages. *Lake Flixton* – as it is known to archaeologists – was a wide, relatively shallow body of water teeming with fauna, fringed with wetland, swamp, thick woodland and seasonal Mesolithic dwellings dotted around its peninsulas and promontories, thought to be used by hunter-gatherers as a seasonal base. The finds have been revelatory – huge quantities of flint-felled timber created a firm platform on the lake's shore; posts for a permanent building with a wooden frame lashed together using honeysuckle stems or nettle cordage and thatched with reeds; hundreds of examples of flint and deer bone worked into bodkins, scrapers, harpoons and arrowheads. But the relics that grab the heart and haunt the eye are the twenty-one headdresses. Fashioned from the frontal bones of male red deer skulls, the antlers still attached, these are frightening, primal things, entrancing and unnerving to behold even when sterile and suspended behind a layer of Perspex at the British Museum. They have a strange presence and an almost human quality. On each, beneath the antlers, just where the plane of the frontlets is flattest, two holes have been neatly bored through the bone using flint. In that peculiar way our minds seek out the form of faces, you invariably see these as bifocal eye sockets, as though you are being confronted with the skull of a small, horned person staring back at you. In fact, the holes were made at the back of these skullcaps so that they could be tied to the head with a leather thong. These weren't hung like hunting trophies today, bleached, mounted on a plaque, nailed to a wall to gather dust; they had a purpose: they were made to be worn. Each was carefully worked on and thinned down in places. Cut marks show that the deer's flesh was carefully removed before the bones of its nose were snapped off and the edges of the skull trimmed. Their insides were scraped and smoothed, made comfortable for specific human head shapes. It was bespoke tailoring. For what, though? Celebrations? Rituals? Shamanistic dances? Rites of passage? Disguises for hunting? The definitive answers aren't preserved in the peat but what the deer bone tools, carved antlers and head-dresses make starkly visible is the extent to which our bodies must have been permanently animalised. They speak of a particularly porous boundary between human and deer. And perhaps it needed to be. The act of hunting was fraught with tension and danger: a broken leg could mean lameness, a gored stomach a slow, painful death. Maybe the further we pushed through the membrane and the more we coaxed the animal within ourselves, the more skilful and successful we became. What the relics of Star Carr also exude is a feeling of the local, something that is so often absent in great archaeological finds. This wasn't something that happened halfway around the world; it was right here. The long reach of time has brought distance, but even so, when you think about it, it'd be surprising if there weren't these kinds of involuntary emotional, physical, spiritual responses whenever the relationship is briefly, unexpectedly restored. It's like old lovers randomly bumping into one another in some far-flung foreign city – strangers who are instantly and intimately familiar.

Proximity's the thing, I'm sure. The sort that you feel when you look into a living eye as it fixes upon yours. Old memories are stirred, and not just from the human side. In that second or less, the deer took in my details too. A flash of remembering, that fear flickering down its flank. Until very recently the idea that memory could be passed on genetically in animals would have got you laughed out of any laboratory. Scientific consensus was adamant the slate was always wiped clean from one generation to the next. Yet recent studies in the field of epigenetics have revealed compelling evidence to the contrary. By training mice to fear a particular smell – cherry blossom, of all things – researchers at the Emory University School of Medicine, Atlanta, have shown that the experiences of one parent influences the structure and function of the nervous system of subsequent generations, meaning that the same stimuli can elicit the same emotional response in animals that have no reason to possess it. If it proves true across the mammal kingdom that environment, experience and traumatic events are sufficiently powerful to take root in DNA and be passed on, who can tell what a deer might recall in the moment it sees a human shape lying in a hollow below. Who really knows what might be inherited and transferred?

*

Seeing one deer is never enough. That's what becomes apparent as I crouch in the hollow, looking west and east along the river, willing the shotguns to flush another. A branch snaps and I turn towards it, craving the sensation again. It's not about blood or killing; it is a yearning for closeness, the visceral flood of animal without and within.

When Mesolithic hunter-gatherers stalked here roe deer colossally outnumbered humans, but by 1800 they had almost been driven to extinction in Britain. Now they are common again in urban and rural areas. Edge-lands, overgrown and largely people-free, provide perfect corridors between terrains and plenty of rough cover to hide, feed and breed in. Two months ago it was as though a pair of does was waiting for me twenty metres or so east of the holloway. An early morning mist cloaked the field. Their heads barely broke its surface. At first they seemed like giant rabbits, ears twitching and turning as they ate. Then a motorbike accelerating hard on a road triggered their flight reflexes and they bounded off in silence, disintegrating through the stubby spikes of hawthorn hedge, like leaf-smoke. Thrilling enough, but it's different when you could have reached out and touched a living deer, when you might have stroked, or stabbed, the flank flexing in front of you. That kind of physical intimacy doesn't feature in our daily lives any more. It must have started to fade as the shift towards settlement and farming took hold, before being denied completely as hunting forests like Knaresborough became the fiercely guarded demesnes of royalty. For most of us today the only connection is through the filter of a screen. The watcher has replaced the hunter. Watching was always a vital part of hunting, of course – following slots imprinted in the ground; noting movement, behaviour, location, weaknesses – but now it is the entirety of our engagement. Wildlife documentaries do astounding things in bringing us closer to species we would never otherwise learn about, taking us on voyages to the deepest parts of the ocean or stretches of remote jungle, revealing sights inconceivable to our ancestors. Even so, they are only ever a passive process, cerebral and reserved; there is a far more complex, complicated and profound wonder to seeing wild animals close-up and in the flesh. Interestingly, though, for all the separation that's occurred, the connection still refuses to leave our systems. The affinity is there in the nature-craving TV schedules, in our fashions, even in the brand names of our cars. Science and technology – fields that at certain points in history almost defined themselves by their distance from, and mastery over, nature – are returning to the source for inspiration. Biomimicry – the study of nature's designs, organisms and ecosystems to solve human problems – is a fascinating and expanding business. I've read of car companies using the scales on butterfly wings as a model for solar panels, and American running-shoe brands creating soles that replicate the traction functions of a mountain goat's hoof. In this emerging area deer are proving their worth to humankind again. Scientists at the University of York are working on replicating the structure of antler as a basis for incredibly tough, resilient materials. They found that on the verge of the rut, just before bucks or stags duel, their antlers dry out. Instead of becoming brittle and breakable as you might imagine, this process actually makes antler two and a half times stronger than wet bone. York was a Mesolithic site itself, probably chosen for its position at the confluence of the Ouse and the Foss rivers. Over the same earth where edges of worked flint once cut deer skin and flesh, scalpels are being drawn and microscopes zooming in, still in the interests of sustaining our species. Or, at the very least, making life more durable.

And as I'm sitting here waiting, another thought hits me. Last week my brother dropped off a boot-load of stuff at our house: hand-me-downs from his kids for when the baby comes. The rich haul included a whole box of books and if I think about it now, every one of them had a different animal or bird on the cover: hare, snake, elephant, lion, panda, penguin, whale, bear, fox, owl, swan, mouse, deer – even the mythical creature on the front of the dog-eared copy of *The Gruffalo* is a composite of wild creatures. Surely it's not a coincidence that the names, shapes and characteristics of wild animals are among the first things we teach our children, or that these are the books they love and lap up most readily. It must go deeper than just exotic colours or shapes on a page. I will do the same with our baby, no doubt, and as soon as it's old enough, I'll bring it down here to the edge-land. We're still conditioning and teaching a process of watching from earliest days; there is still an undeniable affiliation. I wonder if it's because in contrast to our vastly altered existences and increasingly unsure world, wild animals remain relatively unchanged, even if we have changed their environment beyond recognition. The deer that jumped over me was a single animal but it was also a link in a chain, an assertion of place, history and time. My shock and excitement weren't because it was alien; it was the opposite – a half-remembered thing, known, forgotten and recalled. A ghost in the woods and in the genes. What that adds up to I'm not sure, but seeing it felt like closing a distance, scratching an itch from somewhere back down the line.

*

The light is changing now, drifting from its height into a watery green. The guns have ceased their shooting and the woodpigeons are cooing away, consoling each other. Blackbirds are tuning up. I can hear a moorhen on the river, under the far bank of emerald larch. Time to go, but I don't want to. I don't want to put my life back to normal. I'm not ready to lose this sensation, to leave the spot where a window was flung open and a world revealed. So I sit back in the hollow. A watcher, waiting.

ONE DAY

A shimmering horizon, the sky baby blue, and this triangle of edge-land has never looked brighter or richer. The lane and holloway grow wild with cleavers, nettle, purple vetch, knee-tickling grasses and the white lace finery of cow parsley and wild carrot; the meadow is flush with daisies, the gold of buttercup and yellow dandelion. Last night, quick, chaotic storms stirred the murmuring, sleeping town, but by morning they are no more remembered than an uneasy dream. The wet air steams with early sun, kindling the fragrance of the whipped-cream hawthorn blossom dolloped over the many tangled crowns, boughs, hedges and shrubs. It was known as 'May flower' once – sometimes shortened further to 'May' – the only tree to be named after a month, but no one calls it that any more. Not around here, anyway. Opening, ripening, the aroma of the cup-shaped blooms is stifling. Part-honey, part-human musk, it fills the air, the nose, the head, hazing everything with what people used to describe as a 'carnal scent' – the smell of sex. It carries too, drifting into the windows of cars gridlocked on Skipton Road, swirling into suburban kitchens and through the vents of passing commuter carriages, wafting down, deep down, into the insect-buzzing trees of the wood.

Beneath this intoxicating air, gods stir in the river, making for the shallows. Mayflies. Crawling over the algae-rimed stones, the nymphs of *Ephemera danica* are six-armed Vasudhārās whose tusky protuberances look like ornate headdresses in the sun-percolated water. After two years submerged and feasting on dead matter, hundreds, thousands, millions of these thirty millimetre bodhisattvas are emerging from silty burrows, slowly being urged to rise by the bubbles of gas accumulating under their exoskeletons. There is nothing to be done. No turnaround can occur. To survive they must channel their energy now into propelling up into the flow and through the rubbery tension of the water's surface. Those that resist, exhausting their strength by clinging on or stubbornly diving back down for the safety of the sediment, will never experience the higher realms.

One nymph lets go, a female. It releases and drifts tail-high and backwards before turning its cylindrical, segmented shape towards the sun, long trident tail fanning and bucking it through the water like a tiny dolphin. More follow, ribbons darting and twirling up to the light. Trout have been waiting and thrust from channels in sorties to snatch at the exposed nymphs, but with every swirl of their tails they drive more upwards. The larvae begin to punch through the film, climbing into the hot, moist daylight with an almost human look – like someone hauling themselves out of a hole in a frozen lake. This action breaks open their see-through exoskeletons, their *chucks*, and they emerge unfurling upright grey-green wings, forming the silhouettes of graceful star-class sailboats. Their mouthparts have ceased to function now; their death is predetermined and irreversible, governed by the energy reserves built up as a nymph. With bodies the cream of hawthorn blossom, they float on the water changed, new creatures, *subimago*. These are the 'duns' waiting to fly.

At 11:02 a.m. Lauren Jackson finishes the early shift, pushes open the double warehouse doors and unclips her name badge. The air is hot as a hairdryer and smells of tacky tarmac and the heated vegetable contents of the red Biffa bins docked like container ships behind Sainsbury's. She sits on the pavement and rolls a cigarette, looking down the road. Despite her manager's complaints about staff smoking where customers might see them, this is where she comes on breaks and for a post-work smoke. It is a short street with a simple arrangement of low Victorian terraces running down either side, but across its end is a chain-link fence where every vestige of town drops away as though sheared. A dip in the land to the west conceals the sloping maze of houses, allotments, roads and sheltered accommodation, so it appears that the road launches straight into distant fields, hill and sky. A landscape of old England; a Gainsborough behind glass. It gives the feeling of extraordinary freedom, as if you could escape into it at any time, even if you never make that leap.

Lauren looks out at it for a while and then pops a compact. Curling stray L'Oreal Hot Chilli Red hairs behind her ears, she breathes smoke away from the mirror and fixes it over her stunning, rough, brown eyes.

'Fit. That's what you are.'

Joe is walking along the low wall behind her with his shirt off, shoulders already pinking in the sun. Happy, handsome Joe; good-looking, full of life. He grins and Lauren smiles back.

'And you're late.'

'Had to get these, didn't I?' He swings a white plastic bag sagging with cans. 'And it took me ages to buy your present.'

From his pocket he fishes out a pack of Marlboro Gold and throws it to her.

'Fags. Wow.'

'It's inside.'

But she knows this and has already flipped the top. The edge of a little self-sealing transparent bag has been folded to fit among the remaining cigarettes. The smell of the bud is overpowering: burnt popcorn, oil, herbs. *Fox*. She breathes it in.

Joe crouches behind her and slips his arms under her breasts. 'Happy Birthday, babe.'

There is a rendezvous planned with mates twenty minutes later at Lauren's dad's house, but it needs to be quick. Friday is his drinking day, much as he might pretend it isn't. He still goes out in his paint-splattered overalls as though off to work, but when he returns for lunch (12:30ish) he's rarely less than three pints in. And that's just the warm-up. There's no violence in him any more, not like when Lauren's mum first left, but he's deathly quiet and morose and, in his daughter's eyes, it's just as unbearable. She hates it – the giving up, the defeat and the lifelessness. So in her little bedroom she quickly peels off the black trousers and purple polo shirt uniform and wriggles into good under-wear, leggings and a skinny vest, checking the window for his van driving oh-that-teensy-bit-too-slowly up the road. Only one card by her mirror this year, *To My Little Girl*, with 'Lolo! 18!' added shakily in her dad's hand. Soon as she opened it (no need to tear, the envelope seal was still wet), she recognised the crap cartoon font from the display racks in the newsagent's next door. Probably been sitting there for eighteen years.

At the kitchen table downstairs Joe skins up a joint then lifts the back-door catch when three raps sound on the window. Lauren rushes in and sweeps her hand across the table, brushing the baccy and stray Rizlas into the bin. Then they're all out of there, out into the heat of the street, the rattle of Water Board jackhammers, the overgrown dandelions in the yard and concrete dust blown up by passing lorries. There are four of them: Joe and a mate from college; Lauren and her best friend Immy, both wearing their sunglasses like Alice bands.

'Where are we going anyway?' Immy asks, pulling hers down, checking her look in a car window, but Lauren is already gone, threading through the traffic.

There are mayflies everywhere, leaving the slipstream and turning slowly in the eddies in groups of twos and threes. Miniature regattas. More duns drift off with the current under the viaduct and over the heads of the waiting trout facing upstream in the weak-tea water, swimming lazily to keep stationary. Each fish knows the flow of the river and where the channels provide the greatest riches of subimago. Each fish barely shifts a fin, holding its position in the flow, still as a kestrel over a cornfield. Then a tilt, flick, and the *blop* sound as it breaks the water's skin and takes another. Fish gorge themselves until something in the drooping willow and alder boughs, the gold air and the hot, heaped-up grass of the river's edges lures the duns into attempting flight. The mayfly is unique in the animal kingdom as the only creature with two adult winged stages. As it is still sexually immature, the only purpose of the first stage is escape. Suddenly, stretching and beating their wings, the duns begin to leap and lift, careering clumsily into the shelter of vegetation.

A grey wagtail waits and watches on a semi-submerged stone. Breast a bright cadmium yellow, body tapering into the fine point of its long, folded wings and tail, it looks like a horsehair paintbrush halfway through a Van Gogh sun. Flying in a short circle, the bird plucks a few of the airborne forms, then, beak bristling, rests with its tail bouncing like it's counting their numbers. But the weight of duns emerging is too much to monitor; they float up to the bank-side leaves, stems and trailing blades of green. Each lands weightlessly, basking, ripening in the warm threads of sunlight, spiny forelegs bent, wings straight and three tails extended like whiskers. Their final stage is already beginning. Even in apparent stillness, the mayfly never ceases to move; it is always folding in, pushing out, reforming, like the walls of the ever-expanding universe, or the edge of a town.

Lauren is the only one who knows where they're going so she leads, but even if she didn't, she'd probably still be at the front. Working their way down the hot, empty tarmac runways of royal-sounding streets – Albert Road, King Edward's Drive – they come to a back alley hemmed in by the high fences and lines of locked garage doors, where dumped rubbish bags have been split and strewn by foxes. She lights the joint with a sharp inhale and holds it. A few more steps down the runway and *Take Off*. The slow release of excitement in the stomach, the skyward lift and simultaneous sinking inward of the mind, the sudden malleability of tedium and boredom, the potential for it all to become something different, something beautiful and mysterious.

The alley leads into a tatty car park pitted with collapsed asphalt. It is a sump for the houses around it, surrounded by sow thistle, dock, nettle and brambles. Everything is jewelled with litter – a bright pink prawn-cocktail crisp packet, sheets of soggy paper, plastic bottle caps, a rusty shopping trolley coiled with the green heart leaves of bindweed. Joe's mate Nathan, dressed in a black Lonsdale T-shirt, high-tops and jogging bottoms, drops an empty cider can and kicks it ahead of him. They follow its rattle along the track towards a metal railway bridge scrawled with a bulbous graffito and a solitary lamppost dressed in a tutu of barbed wire. Running alongside is a galvanised steel palisade fence, the top of its metal points split and peeled like bananas to heighten the treachery of its cutting edges. Beyond it lie the last few houses of red northern brick and a rectangle of yellow: an enclosed patch of waste ground wild with ragwort and dandelion flowers. A collapsed sofa slouches at its centre, its exposed, fat-like cushion foam colonised by invertebrates. Leaving it all behind, Lauren registers the shift towards a place beyond restrictions, out of the way of town, out of the way of people. A lightness somewhere between her eyes.

At the intersection of the old railway and Bilton Lane, a completely different vision: a spectrum of greens to thrill her now slow-blinking, dilated eyes. Near greens and far greens, lime greens and greens that make her think of the beer-stained pool tables at High Harrogate WMC. This place is ablaze with life, though, not stale with slow decay. Every cranny and fissure is filled with wildflowers she doesn't know the names of; there is the musty hot-skin scent of Joe's burning torso in the warm air. The sun is so bright it falls like a cape of gold on Immy's bare shoulders. Another joint is rolled and passed around. Lauren takes in the last of the tangy, tarry tail end and stares down into the verges, entranced by the powder-blue flowers of forget-me-nots and the hairiness of the sticky stems of cleavers. In the heat of midday, she imagines she's melting into the old, ivy toadflax-coated wall they're leaning against. She can hear the timeless vibrations of the million worker bees and, far off somewhere, council lawnmowers trimming verges.

Laughter. Nathan has one arm around Immy's waist, fingers in the back pocket of her jeans, and they slurp from a fresh Strongbow. Minds are slowed, senses paradoxically dulled and thrown open. They stare at a piebald horse and its nervous foal at the bottom of a sloping field.

'D'you dare me to ride him?' Nathan says. 'I fucking could, you know.'

'Which one? The baby?'

'*Shut up . . .*'

More laughter. More bravado and flirtation. Four-to-the-floor beats tripping tinnily from Immy's iPhone. Nathan's lighter flicked to touch cigarettes. Heavy-handed acts designed to show off a sort of tough kindness, all rehearsed and perfected of course. Joe laughs at him and suggests that they skin up in the meadow, but Lauren shakes her head, pushes off the wall and crosses over the old railway, ducking down under hawthorns where its scent is heaviest. Thick as department-store perfume counters. 'It's this way,' she says.

Nathan groans. 'Where now?' He sounds weary, too stoned.

Lauren doesn't break step but points ahead, up the lane, dark and cool with canopy shadow, and off over the scrubby fields and wood.

Joe peers down after her. 'You feeling all right, babe?'

'Course. But I'm going further.'

'Can't we just chill here?' he says. 'There's no fucker about.'

'No. Not here.'

'Where then?'

'The river.'

Now they're confused. The others didn't even know there was a river near town. Mutterings. More groans from Nathan.

Lauren looks back, narrowing her eyes, standing bold, fierce and beautiful. Hera with a crown of hawthorns. 'Trust me,' she says. 'You only live once.'

And they do.

By now the *Ephemera danica*, those Vasudhārās, are awake and fully formed. Long, glossy, crème-caramel abdomens, segmented and intricate, have the kind of wispy brown tobacco smears once found on old magnolia pub walls. Elevated above their previous aqueous universe, poised on the alder leaves, cushions of wood ear mushroom and pole-like grass stems, the mayflies took little over an hour to achieve their ultimate incarnation, to moult into the sexually mature 'spinner', the *imago*. They appear more clearly defined and sharper, as though an aeronautical engineer has stepped in to improve their designs, readying them for their last, triumphant function. The six-jointed forelegs stretch further than before, the three-pronged tails whip out from their rears for better aerial balance and the wings have lost their fine hairs, becoming translucent and etched with black veins. On wider bank-side leaves sometimes two or more of these spinners sit side-by-side. Then small differences in appearance become apparent: the males are smaller in size and darker, with larger, pronounced eyes.

Time is of the essence and yet there is no sense of time. Not as we know it. No fear of the coming, inevitable unknown; these are prehistoric creatures of the present, 300 million years in the making. An order older than dinosaurs. Time to them is in the frequencies of the surrounding bird-song, the fluttering of wings, the sun moving through the foliage, the colours that move across their compound eyes, the vibrations that spill down from a passing heron's croak. Light spills down too, a hot afternoon light that fractures the wood, falling in shards between trees and water. The infinite motion of the river runs in one direction; the endless flux of sky meeting wood in another, and into this strange dimension, as though an irresistible force possesses them, the spinners rise on stained-glass wings, like angels.

Her dad used to call it 'Duffer's Fortnight', this spell when the mayflies were up. When she asked him why, he explained it was because no one ('not even a blind bastard dipping a broomstick') could fail to catch the trout when they were snapping away at whatever floated past. 'It's practically suicide,' he said.

It was always about now, a sunny day around the time of her birthday, when he'd go off to his lock-up and dig out his cane rod. Then they'd sneak down here together, provisions packed into the mouldy knapsack he kept from his army days, pop and sweets for her, Skol and Regal King Size for him, both keeping an eye out for anyone who might ask for a licence or angling club membership. He never owned either of course; then again, they never saw another person down here. He said it was because of the sewage farm around the bend of the river, said people didn't want to fish too close to it. But he swore it was safe, claimed no mayfly would breed in dirty water. *Sensitive souls*, he called them.

'And they only live for a day, Lolo,' he once said to her, catching one in a fist and holding it out to her. She looked at it, wing crumpled, still trying to lift itself off his palm, clawing. 'Imagine that. *One fucking day.*'

And she had.

'So you gotta be quick, right? Seize the moment.'

Then he'd turned, tied on a fly and fed out the line, swishing it back in a looping arc until it became indistinct in the insect-clouded air.

That was then. Back before the flashbacks, the divorce and the hard drinking. Back before he chose a long, slow, selfish death in front of her eyes. Back when Mum was still at home and they kept a scrapbook of found things together like crow feathers, dog-rose petals, once even a four-leafed clover to press between the pages. That was when such things mattered, when everything seemed alive. Then boom. Before you know it, all of it gone.

Except now, just like then, Lauren sits on the riverbank. Eyes full of wonder, drawn by reasons unknown to her, she watches the mayfly's brief, beautiful dance.

The male spinners collect into loose, drifting clouds a few feet above the water. There's safety in shoaling like starlings, like sardines. At first glance, or when seen grey against the sky, they appear as smoke, behaving in the same shifting, slipping way a column from a bonfire does when blown across a motorway. Then they roll into tighter, tornado-like vortices, sometimes visible, sometimes lost against the leaves. In this way countless mayfly bounce around this stretch of river, floating, climbing, falling, passing over the drowned branches and the moorhen nests. Nearer, they look more like the blizzard of dust motes you get after wheat has been cut or glowing dandelion seeds caught in sunlight and a soft breeze. And yet for all their wild, mad dancing, these male mayflies never touch. All moves are planned in this ritual. Contact is reserved for when the female spinners circling on the peripheries dive suddenly into the columns' centres. Using elongated forelegs, the males intercept them, grappling the females by the thorax in a mid-air embrace, mating with her and then releasing her so quickly that the human eye can barely perceive the coupling. A moment of pure life, lost in the veil of the swarm.

They love it, of course, this freedom. She knew they would. After daring each other for a good hour, Joe and Nathan strip down to boxers and tippy-toe over the rocks, mocking each other's flailing, pitching walk. Screams and laughter. Now they splash about its deeper channels, showing off, shouting and laughing, throwing handfuls of mud then suddenly losing their balance and drifting before dragging themselves into the shallower riffles again.

Lauren joins Immy beneath a willow tree where the grass bank becomes the clay-coloured sand of a small, crescent beach. Below, bags of Strongbow cool in eddies overhung with trees. Strewn over the sand are the lads' clothes. Chewing chuddy, Immy sits cross-legged, a can clasped between her denim cut-offs.

'Here.' She hands over a half-smoked joint and then pulls her T-shirt over her head revealing her flabby torso and pink plastic belly-button stud.

Lauren takes the smoke, fills her lungs and peels off her

vest too. Both lie together in their bras with the sun full on
their skin. It feels like a gentler heat now, older, that
getting-towards-evening sun. Long draws on the joint as
Immy whispers to her and pops her gum. 'Do you think
Nathan fancies me or what? I really want to lose some
weight. I'd give anything for your figure, y'know...' Lauren
listens and comforts, but it all sounds so distant, like she's
higher, up there with the clouds of mayfly, playing the same
game she used to – trying to follow them in the air to see
where they go. Circling and circling.

Neither of them hears Nathan until he is walking up the
bank with his clothes in his hands. He looks unsure of what
to say.

'Do you know what's down there then?' he grunts eventually, jerking his head off towards the bend in the river
downstream.

Lauren shakes her head.

'Why?' says Immy, shielding her eyes, sitting up. 'What's
up, babe?'

'Nothing. I dunno.' He shrugs and reaches down for
a drink. 'Thought I might go look. Thought it might be a
laugh. You wanna come?'

Immy looks at Lauren, then back at him. 'Yeah, all right.
I'm up for a laugh.'

As Nathan walks off to get dressed, Immy hangs back.
Then, when he's out of earshot. 'So I might see you back
in town then. Is that all right?'

'Course.'

'Thanks, Lo. Oh and...' a kiss on her head, 'Happy
Birthday.'

Lauren watches them walk away, Nathan slipping his
arm around Immy's bare midriff as they disappear into
the greenery.

'So you coming in or what?' shouts Joe from the water. He's doing backstroke.

Lauren laughs. 'Better idea. Why don't you come out?'

Joe doesn't know how good he looks rising from the water, his muscular pale form emerging through the peaty brown. Happy, handsome Joe; good-looking, full of life. When he wades out, his broad, burned shoulders turn white in the light, his hair is plastered over his forehead. Mayflies land on the golden-haloed outline of his head and he doesn't even realise. Lauren grins and he smiles back. It's then that she sees him as something more – a shaft of light, part of the million beautiful growing things all around her.

'Why've we never come down here before?' he says, dripping wet, skin goose-pimpling as he stands above her wiping his chest with his shirt.

'We're here now, aren't we?' she says.

And she means it. She means we are here, now. Just us. There's nothing else.

Somewhere back there is a world of financial storms and wars, a world of shitty shift patterns, rotas and customer service training. Tomorrow's early-morning stock-take. The 6 a.m. start. The collective denial. But that's not *real* life, not like this. Not like this one perfect day.

One fucking day, Lolo! So you gotta be quick, right? Seize the moment.

She lifts her hips and rolls down her leggings, then lies back on her elbows. Joe seems suddenly shy, intimidated by the perfection beside him – her curved, caramel body, the black satin bra and knickers. So she grasps his arm and kisses him. His mouth is still river-cold. His flesh washed with wild water, but it only makes her want him more and she pulls him down onto her, down into the seeding rye grass. Then everything becomes details. The swell of his bicep and the nape of his neck. The smell of skin and saliva. They way they kiss in-between undressing each other, the sun still warm on their bodies. His lips along her collar-bone. Her legs wrapped around his back. Sweat on her reddening chest. Open eyes, closed eyes. Foreheads pressed together. All the time so quiet, so intense.

Afterwards, Joe rolls onto his side, his arm across her, watching her face. Gravity anchors Lauren to the earth. She focuses on the rise of her abdomen when breathing in, then the stream of warmth through her nostrils as she exhales. Nothing happens next. There is no destination. This is it. Here and now. And as she lays there, her shock-red hair tangled with purple loosestrife, she senses a sudden emptiness. Above, the mayfly are fading from the air.

The flurries of ecstasy wane with the dying sun. Plumes break up and drift apart as, gradually, spent male spinners crash down to the water or, more commonly, the grass and leaves of the river edges. No one knows why they might be drawn to the land to ebb away their last, but by nightfall they will decorate the silk of spider webs and fill the bellies of finches, moorhens, frogs and bats. All the while, just beyond the battery-running-out flickers of weakening wings and forelegs, a peculiar green and yellow evening light shimmers on the Nidd. Here, on this unloved stretch, where Celts once sank votive offerings in the hope of raising nymphs and naiads from below, the female mayflies return to the water to do the same.

Keeping about a foot above its skin, they head upstream. Often flying in circles, each wafts down on rearing wings, curling its long body, dipping its abdomen into the shining screen and releasing a batch of fertilised eggs. Sometimes

the spinners settle for a few seconds on the surface, sometimes they barely appear to touch, but every time they meet
a reflection of themselves rising from the deep. This strange,
driven, determined action is taken without thought of the
risk from feasting trout and, now, the mallards and duck-
lings that are leaping and flapping for them. Many are
taken this way, many more survive to reproduce. A thousand eggs here, a thousand eggs there – up to 8,000 per
spinner deposited in intervals – creating final ectoplasmic
clouds in the water below, each a swarm of millimetre-wide
eggs that sink to the bottom and attach to the rocks, sediment and weed. In ten days, these will hatch into little river
gods, burrowing into the sediment and beginning the cycle
again.

Time flies. *Timeflies.* It could be a name for the mayfly.
What seemed an unstoppable orgy of life only an hour ago
has receded like a disappearing universe. The air has turned
everything copper and bronze. With all their eggs and
energy expended, the female spinners tire and fall to the
river's surface. It's now that they look oddly human again:
either confused by the tension holding them or embracing
it, laying their wings across the water as if, with purpose
fulfilled, they no longer fear death in the jaws downstream.
And like this they drift on, a million crashed gliders. Falling
quietly around them, the rusted hawthorn blossom. Before
long there will be no trace that either ever existed.

THE TURNING TIME

It was just after dawn when I broke into Bachelor Gardens Sewage Works. To my surprise, no alarms rang. There were no security guards, no flashing lights or sirens cutting through the muggy air. You'd think that in our surveillance-swept world I'd have triggered something, somewhere. At least, I thought so; so I waited a while, my legs crawling and burning with nettle stings, bracing myself for the inevitable detection, rehearsing the words to talk my way out of arrest. Minutes passed in half-light; the pain intensified; nothing happened. The metal doors in the spectral floodlit buildings and sheds remained shut. No one yelled at me from the shadowy tanks or the gangways running between rust-streaked vats and water channels. There was just that simultaneous hum and whine of machinery I'd heard so many times from the other side of the fence, now louder and accompanied by the muddy brew of human waste and chemicals rising from filtration beds.

My thoughts turned to escape, for this was not a deliberate act – who in their right mind breaks into a sewage farm? Rather it was the result of a wandering mind and some (in hindsight) ill-advised off-the-beaten-track running. The baby kicking and turning inside Rosie had been giving her restless nights. I'd risen early to let her spread out across the bed and to try to shake off the shackles of too-little sleep. Approaching the edge-land from a new direction, east, through an unexplored labyrinth of cul-de-sacs and estate roads, I chanced on an unruly doorway of tree and shrub leading down to a little river. I ignored a path looping back and, instead, plunged on, running along the water's side through chest-high vegetation to where I hoped the beck might link up with the Nidd further on. Then, suddenly, my legs were on fire. Nettles concealed in the hogweed and jungle-dense Himalayan balsam had ambushed me. Flailing, flaying, angry leaves waited for the slightest movement to inflict new wounds, but the closest shore in this sea of stings lay ahead and I leaped for it like a triple jumper, hardly noticing the ramshackle wall or the collapsed curtain of fence and old wire. Hardly noticing, that is, until I'd climbed over and pushed through and by then it was too late anyway. I was trapped in this otherworldly facility.

No one was coming. That was clear. So I searched for another way out, one that might spare my legs a second lashing. The perimeter was locked-down; all high walls and sturdy mesh topped with barbed wire. The main gate was chained and its sign graffiti-scrawled: 'The Smokers Yard', sprayed in blood-red paint with a scattering of roach ends and cider cans to illustrate the point. Almost willing to be discovered, I headed back to where I'd broken in, but via a different route, a small road that wound between the conglomerations of barrack-style buildings and choppy brown lagoons, past surreally idyllic stands of Scots pine and triangles of mown lawn. It was all very Cold War, the air a post-apocalyptic purple and pink, warm, misty and poisonous-smelling. Everything was deserted, eerie and, if

I'm honest, a little exciting. Up, over a rise, and through another screen of pine trees, I was suddenly confronted with a strange geometric pattern sunken into the grass: six large, wide, water-filled concrete circles. Reaching out from their centres were great metal arms that spun slowly, churning and stirring, like the exposed mechanisms of some ancient buried machine. I could smell pine resin, freshly cut grass and sulphur. Beyond lay the huge, hanging silence of the Nidd, the gorge, the viaduct, the wood and, further still, hazed fields of rapeseed and wheat. But it was there that I saw them; there, in the air above those stinking ring-pools that they circled, silently, secretly.

There must have been eighty, maybe a hundred, but so many that at first I took them not for birds but insects clouding in whirls over the drums, turning the way tea leaves do when washed down a plughole. I forgot the pain in my legs and stood there looking east, as the rising sun laid its hot hand across my forehead, transfixed by their flight and strange flocking, feeding formations. There was none of their usual velocity; the birds drifted in the air as if in slow motion, slow enough that I could make out each one's profile: the stiff-winged, black anchor shape against the sky, reversing direction at will, taking what must have been millions of the stirred-up, snub-nosed sewage flies in balletic sweeps and dips. Here was the tidal wave of summer in its infancy, still out at sea, gathering strength. My boat had somehow drifted into its path. After a while, standing there staring seemed voyeuristic, as though I'd caught them early, backstage, doing warm-ups before the full show. I felt that I should say something, maybe cough politely, let them know I was there at least. But what do you say to birds?

Alarms, security guards, police sirens – they're what you might expect when you break into a sewage works, not a sky full of swifts. Such things lift the lead from your head. Worth crossing a sea of nettles for; twice, as it turned out.

Where they went next, though, I couldn't tell you. Those last warm days of May broke, and in swept the storm-horses of rain and wind, catching the country out in a stampede, kicking down the flowery frills and thick green bunting with heavy, iron shoes. The light changed. Skies slated. The outside intruded. Slugs invaded the kitchen every night, leaving silvery ghostly ribbons all over the floor. Weather bulletins showed a graphic of an atmospheric depression migrating east across the Atlantic, settling over the British Isles. Its isobars corkscrewed in a lazy circle, a scribbling motion round and round, like a bored child's crayon. 'Here for the next month,' the man said. And it was, all day, every day, from dawn to dusk until the Nidd surged high, loud and dirty brown. The woods and the wheat fields shook and cowered like slaves under an overseer's whip. I thought of the swifts often, but each time I donned a cagoule and ran down to the sewage works, seeing anything was impossible, like the sky had fallen in and lay bubbling on the earth. Everything was rank with a sulphurous fog. Eventually I stopped bothering. I knew they weren't there. Flies spasmed in soaking cobwebs between the perimeter's barbed wire; I could hear the mechanical arms turning and whining. The egg stink. That was all.

Swifts can't feed in rain and so will travel astonishing distances to avoid bad weather and find food. They are fantastic meteorologists, capable of detecting the finest fluctuations in air pressure and moisture. Then, despite only weighing the same as a bar of Dairy Milk, they will fly into a headwind to reach more clement climes on the fringes of weather patterns. I reasoned this was what my swifts had done. But exactly where they had gone was anyone's guess. Ornithological surveys have tracked swifts leaving gathering points above London to conduct foraging trips over the Norfolk and Lincolnshire coasts, way out across the North Sea, even as far as Germany, to plunder the abundant clouds of insects that swarm at the rear of an occluded front.

To grounded minds, feeding excursions that might clock up 600 or more miles in a day seem incomprehensible, inefficient even, but the more you learn about this bird, the more you realise that considering any part of its existence by our limits and measures is a mistake. Swifts are almost entirely beings of the air, as near to an element as you'll find in a creature, evolved to spend their lives in a state of permanent airborne motion. From the point that they fledge and free-fall from nests on those sickle wings, the swift's life is one long aerial journey. Unless injured, they will never touch the ground. They feed, drink, preen, mate, even sleep in the air. Only fleetingly when nesting high up in the nooks and crannies of old buildings do they become creatures of the lower realms, of our earthly world.

Just as the storms had sent the swifts soaring, the dreary, pelting days drove me inside. Each morning through a curtain of grey I waved off Rosie as she set out on her own swift-like excursions across the county. Since moving north she'd taken a job selling produce into delis and farm shops, which necessitated considerable road miles. Her 'territory of responsibility' stretched as far as Lincolnshire in the south and Whitley Bay in the north and took in pretty much every back road in-between. Working long hours at my desk as rain machine-gunned the roof tiles above, I worried about her out there, driving alone under dark, unrelenting skies. Soon flood bulletins began to pour in by the hour. The ground was saturated. Reporters dressed in that uncomfortable mix of waterproofs and ties described how whole towns were being cut off or split in half by bursting rivers. There was talk of climate change and blame, how all of us need to get used to living with these kinds of extreme meteorological outbursts from now on, as if the weather was some moody teenager tantruming through a house. The actual source of the misery took many forms depending on who you listened to: either it was Arctic sea ice knocking the jet stream off course, pushing it south, or it was our overheating atmosphere creating drier air capable of holding more moisture, hence the heavier, increased rains. Whatever the scientific argument being put forward, all seemed to have one depressing area of common ground: the hand of man.

Every day Rosie said the same: 'I'm fine,' but I could see she was exhausted. By the end of the month, she was done-in by the hypnotic motion of the windscreen wipers and the effort of concentrating through clouds of road spray as she followed diversions. One night she fell into a deep sleep almost as soon as she walked through the door. Lying on our bed as the rain squalled against the window, she gently held her bump, our baby, in its own little watery world. I stretched out next to her, slipped an arm under her shoulder and stared up at the ceiling. It was 23 June. The longest day of the year had already passed, unmarked, lost in the flood; the warmth and sun it normally promised seemed further away than ever. I thought about the swifts again, birds that were supposed to be the winged heralds of our high summer and blazing days, and wondered, *Had they come too early? Had they bailed on us? Do swifts make mistakes?*

I wasn't sure, but something was troubling me. Their absence, like the unceasing rain, made me anxious, as though the world wasn't working properly. 'Generally unsettled,' the forecasters might have said.

You'll read of the common swift (*Apus apus*) as a 'British bird', but this is something of a misnomer – a bit like saying the passing clouds are British, or the constellations. It's true that many make the nearest thing to a home – their nests – here, but even the swifts that converge in our skies each year only spend a maximum of three months (usually between May and August) in the UK, a mere quarter of their lives. Really, we borrow them at best. The rest of the time they are in transit or hoovering up insects above the rainforests and rivers of the Democratic Republic of Congo. All the world's population of common swifts overwinters there, living the same fluid, ranging life under African skies. Just as they will in more northerly latitudes, swifts cover huge areas in the search for food. Flying high and fast they will travel as far as Mozambique in the east, Angola in the west, and down to South Africa. Then, around April, with their breeding season approaching, they surge back in their millions, rising and heading pole-wards, undertaking epic and perilous migrations over vast oceans, mountains and deserts to often long-established nesting sites across Europe and Western and Central Asia.

'Our' swifts, as much as we can ever really call them that, were once believed to fly directly north, following what seems the straightest and shortest possible route back to Britain. That's what land-locked logic assumed. However, recent geo-location data from the British Trust for Ornithology has revealed another story. One tagged bird's 'flyway' – as the experts snazzily term migration routes – was found to follow the Congo river west, heading out from its mouth across the Atlantic before turning up in a curve to Liberia in west Africa, taking advantage of feeding sites and wind patterns. There it circled for ten days, feasting and fattening on swarms of flying termites, before a rapid, non-stop flit back to Cambridgeshire across the Sahara, over Spain and France, covering 3,100 miles in just five days. Amazingly, within three months, just as autumn began to cool the far horizon, it made a similar-length journey in reverse. In total, the bird was recorded as flying a round trip of 12,400 miles to breed in the UK. And this was by no means the record – other tagged swifts in the same migratory loop had more than 17,000 miles under their wings.

And to think that these birds make these pilgrimages once a year, every year, cruising at seventy miles per hour as loftily as 10,000 feet, sometimes higher. People claim to have seen them at nearly double that altitude, cruising above the peaks of the snow-crusted Himalayas. That zooms out the mind instantly. Right out and up into the cold clarity of the higher, quieter realms. You enter a kind of Google Earth world where cities are reduced to smudges of grey; forests and field networks are little more than pixels of green. The more you think of it, the more the head spins, as though you're flying up there yourself amid the isotherms and the jet stream, rising with air currents as sun-edged horizons, entire countries and vast, grey, hostile seas spin and vanish between the breaks in clouds below. It's a dizzying perspective. The mind struggles to conceive of how these small, almost weightless sylphs, woven of little more than feather and thin bone, are capable of such ludicrous speeds, heights and distances. And they are nothing short of ludicrous.

I read somewhere once that a swift chick ringed in Switzerland was found dead as it returned to the same nest site twenty-one years later. Observers reckoned that in the intervening years it had clocked up around three million miles.

But for birds migration is not a choice, an urge to travel like ours, born of being too long settled; it is an ancient, hard-wired instinct driven by those two biological imperatives: survival and reproduction. To the swift a British summer is supposed to provide the riches of insect hatches filling warm air and a relatively calm climate for raising chicks. If swifts had such things, the travel brochure for the UK would promise: 'long balmy days and sunny evenings for extended feeding sorties – perfect when there are new mouths to feed!' So what happens when they arrive to a washout? Are those ancient, miraculous journeys all in vain? Almost certainly. Experts tell us that adult swifts toiling in the wet end up woefully underweight and at risk of lacking the reserves to complete the return flight to Africa. Even when they try to nest, scratching the undeniable biological itch, it seems they have an awareness of the hopelessness of it all. Knowing there isn't enough food to feed themselves, let alone young mouths, they abort, pushing eggs unhatched from nests. Over time and given persistently grim summers, it's feasible that swifts could cease to exist at all in our skies. We're only too aware nowadays that species will vanish, but the thought of losing swifts terrifies me. 'They've made it again,' wrote Ted Hughes, 'which means the globe's still working.' How perfect that is. Nothing speaks of this planet's inter-connectedness like the swifts' migration; nothing screams so loudly of its fragility either.

*

The room had grown dark. Tiredness was finally dragging down my somersaulting mind. My head sank back, further into the pillow, descending into the quiet, loamy blackness of sleep. Then it registered something, an alteration occurring in the atmosphere. The rain was stopping. I looked up at the ceiling and imagined the roof peeling away, the clouds dissipating and seeing swifts up there somewhere, swimming broad-shouldered among the stars.

My arm had gone to sleep. I slid it from under Rosie's shoulder. 'You OK?' she whispered, stirring. 'What are you thinking about?'

'Migration.'

She smiled and turned over. 'Please don't,' she said, guiding my hand to her belly and holding it there. 'Stay here with us.'

I laughed and then it dawned on me. *That* was it! That was what I wanted to say to the swifts, what I should have said when our paths crossed at the sewage works. *Stay here with us.* And it was those words I was thinking of the next morning when, opening the curtains, I glanced up and finally saw a scattered band of them flying high in blue sky, east to west, like stray eyelashes blown across coloured paper. I pressed my face against the glass and then wanted to tell Rosie, but she had rushed out an hour earlier to a midwife appointment almost forgotten in the mad monsoons of the last month. Oh, well. A nice surprise for later. Then – what timing – she called. I was on the edge of relaying the news when my brain registered the formality of her tone, the forced calm, the words: *I'm on my way to the hospital. There is...there might be...something wrong.*

*

Hope is such a useless emotion, but it's the default setting in all of us. You can't help resorting to it. And I *hoped* against hope that this was just normal pregnancy gremlins, a regular nothing, some common mix-up. After all, she'd corrected herself; she'd said *might*, hadn't she? With the phone still clasped to my ear, brain whirring, I realised I'd been staring at the silicon seal that wraps around our double-glazing for I don't know how long, noticing for the first time how it had a few little black specks of mould spreading from the corner. I noticed other things too. The heat and brightness of the morning. The softness of the carpet between my toes. *The carpet* – we'd deliberated in the shop and then plumped for the softest and most stain-resistant one, even though it cost more, as we both envisaged little hands and feet crawling over it. All the time, in my mind, cogs were turning, processing. To speed things up, Rosie was driving herself straight from the doctors' surgery and was already near the hospital when she'd called. *Don't come*, she'd said. *Honestly. By the time you get here they will have finished the scan.* The scan? How long does that take? *They put me on a machine for twenty minutes. I'll call as soon as I can. Promise. Don't come.*

The machine is actually a belt consisting of two sensors which is strapped over the womb to monitor the baby's heartbeat. As I washed quickly, dressed and tightened the belt of my jeans, I thought of the same thing happening a few miles down the road and felt sick. I sat down. Over and over a question: *How can I fix this?* I got up again. Making coffee I knew I wasn't going to touch, I replayed internally what Rosie had told me, looking for solutions: the midwife had been going through the routines as normal, asking the questions, all cheery, motherly, brassy, comforting. 'So, what about this rain we've been having?' Her eyes had taken on a glazed, middle-distance look as she 'had a feel' around Rosie's bump and took her blood pressure – quite high. 'Have you been overdoing it? You should be taking it easy at this stage.' Then she drew closer, leaning forward with the stethoscope. The cold disc caught a high, tripping heartbeat first, going far too fast for itself. Odd. So she reset and tried again. And again. Now it was even more unusual, a yawning silence. An absence. It was then she called the Antenatal Clinic.

Every room in our house suddenly seemed devoid of air. I went outside and – wouldn't you just know it? – the day was spectacular. The month had got over its tantrum. Light poured down the street. The mottled stone chimney stacks reached up into a tropical ocean sky, casting long shadows over the slates. You could smell the heat building. Cars gleamed. In front of our door, before the low stone wall and the pavement, weeds had run riot with recent over-watering and colonised a small strip of pebbles completely: pink-flowered herb Robert, willowherb, dandelion, alkanet, the rough leaves of a flowering currant dug up long before our time, yet still clinging on. A laurel bush I'd cut back a few weeks earlier had started re-growing clumps of soft, waxy leaves from its cut branches. I noticed all this: the way nature was defying the obstacles, using them even, the way the laurel had tangled with the fence to gain better traction and strength, the way the currant had forested the pebbles with young shoots. I noticed it all. And hoped. *Could these be signs?* I thought about running to the edge-land, running to the hospital. Then, suddenly woozy, I crouched down with my back against the house. An orb spider's web was knitted perfectly between the laurel's leaves. I fought the urge to project everything onto that spider, but my brain wasn't behaving – *If it moves left everything will be OK. Move left. Please.* I checked my phone's screen, bewildered by how slowly the minutes were passing. The spider sat still. The stone in my stomach grew heavier. And over and over, those four words in my head: *Stay here with us.*

The sound was distant at first, like that note you hear sometimes when still half-asleep with your head buried in a pillow and you breathe out through your nose. A soft, airy whistle. Then it swelled quickly until it was a loud banshee scream – *seeeeee-seeeee, seeeeee-seeeeee* – zipping fast and low over the terraces and the back-to-backs. Swifts tipped into the streets in riots of joy and noise, like a carnival hitting town. Their scream wasn't menacing or melancholy, more playful than anything; the peak of a laughing fit, a toddler being tickled. *Seeeeee-seeeee, seeeeee-seeeeee,* they implored in their glee. And I tried, but before I could even turn my head they were off again, arrowing over another roof.

'Did you see the sparrows? They've gone mad!' said the little girl from next door. I hadn't noticed her coming up the pavement but she stood on the other side of our fence now, foot on her scooter, all red dress and big smiles. She's a kind, confident girl, Libby. Could be a kid's cartoon heroine – a wonderfully bolshie tomboy, bright as sun on snow and the first person who spoke to me on our road after we moved in. Shielding her eyes with a hand on her thick-rimmed glasses, she craned her neck and scanned the empty sky.

'They're swifts, Libby,' I said, 'not sparrows.'

She looked at me, frowned, and repeated the word to herself under her breath. *Swifts.* 'And where do *swifts* come from?'

'Africa.'

Africa, she mouthed it silently again. 'Why are they here then?'

The words were there somewhere, but they suddenly refused to form in my mouth. I coughed and checked my phone. Only a minute had passed. Ridiculous. Then from somewhere beyond the end of the road, from the direction of the edge-land, that swelling sound again: *Seeeeee-seeeee, seeeeee-seeeeee.*

'Here, look. They're coming back,' I said, pointing up at the long blue corridor above, formed by the rows of terraces. That's the trick to watching them, to look at the sky rather than try to follow the bird. And that time we did *see-see*. Immediately three screamed into view and split formation, the two outer birds peeling left and right over roofs; the central swift plunging down into the canyon of houses, banking and brushing up against the side of number 28. It was a quick movement, a deliberate crash: the swift touching the point where wall met the carved wooden roof trim, before turning and bursting off again. It passed so close to us I could see the sun through its tail and hear the wing whistle. Libby put her hands up to her head and whooped. 'It went through my hair!' Not quite, but that's how it feels when they explode past you – as though their seesaw wings are brushing your face, coating it with sky dust and cloud wool.

It's one of the things I love most about swifts – that they manage to pull off part elemental wonder and part town-bird with such aplomb; they are creatures with a truly panoptic world view, possessed of a higher consciousness tuned in to the great governing forces of our globe, and yet, for a short time, birds of the common people too. When they return to our realm it's not to pristine reserves or protected wetlands, it's *here*, to the sewage treatment plants, factories and back streets. The sprawl is as much part of the swift's world as the stratosphere. They're like little, wild, black boomerangs hurled down from the heavens, sent to us to bestow a kind of day-to-day benevolence. One minute you're fumbling with the keys, emptying bins or waiting desperately for a phone call, the next they're with you, above you, around you, a dose of the sublime where you least expect it; when you most need it.

More came, spinning and thwacking into the eaves or bouncing off over our heads. Gangs of them casing out the joint – numbers 25, 20, 13, 12, all the houses soon had visitors. We ticked them off together each time the birds appeared in their scattering, lunatic, overlapping shrieks. I had no idea our street would be such prime real estate, but it made sense – a row of Victorian terraces that, for all its modern UPVC double-glazing and brightly painted front doors, is still decidedly higgledy-piggledy in places. Swifts have a nose for lapsed DIY duties: the fallen-away mortar between sandstone, the inviting vistas between slipped roof slates, the gaps where gable ends and eaves have warped a bit over the years. This is because, over time, swifts have warped a bit too. At some point along the evolutionary high wire they threw their lot in with us, largely abandoning the nesting habitats they evolved with, like tree holes and cliff fissures, for the nooks and crannies of towns. Some have suggested it may have coincided with when those master stonemasons, the Romans, were spreading through Europe. And that would have been a clever move. Beautifully opportunistic. Why wouldn't you ally with an ever-expanding, building-obsessed species with a tendency to leave holes in its many roofs and towers? But what may have been a successful strategy for two thousand years has been unravelling in more recent decades. Swift numbers have been hit hard by post-war building methods and materials; the sealing up of structures in the name of energy efficiency, and a general squeamishness at sharing our homes with the natural world. They don't give up easily, though. Being dependent on the man-made these days, swifts keep looking for the gaps, forgiving our cooling affections and indifference, blessing us with their presence. I love that about them too.

That's what all the wall-bouncing was about. It is known as 'knocking' or 'banging' and it's what swifts do to scope out their nest sites or challenge for occupied spaces. The urgency was understandable. With mating delayed by the weather they were wasting no time with formalities, but slamming about, ringing doorbells one after another to see if anyone was home. Breeding swifts are traditionalists and will keep the same partner and nesting site year on year, provided man and nature are complicit. When arriving at their nest sites separately, the first bird in a mating pair will reclaim the spot by screaming their presence, folding in those broad wings and vanishing into the eaves. Should a rival already be inside they will fight in violent, close-quarter duels, like handcuffed wrestlers, hissing, grappling and scratching sometimes for hours at a time before one is ejected and the victor roosts to guard it and wait for their mate. Reunited, both male and female birds take shifts to gather nesting materials from the air – feathers, thistle-down, leaves, grass, seed cases, even the cotton centre of cigarette filters – all taken on the wing and formed into loving, soft, cup-shaped bowls bound together by saliva. Spit-welded, you might say. The younger, non-breeding swifts – those birds under two or three years – also play the knocking game, usually receiving short shrift from a shriek-ing breeding occupant. Unperturbed, though, they pair up and range out to the edges of a colony, looking for space. In some instances they'll even build dummy nests, practis-ing for when their own time comes. Given their speed it's hard to know for certain, but I'm pretty sure both breeding and non-breeding swifts were working their way down our road. It made me wonder if perhaps our street formed the outer limit of an established swift territory, if our house might not be a swift edge-land.

My thoughts tumbled like this as we watched them, Libby and me. She was a little human blessing amid the avian ones. I was thankful she was there, forcing me to keep my eyes skywards, calming my trip-hammer heart, distracting me with her never-ending list of questions: 'So how far *is* three million miles?'

'Well...it's like flying to the moon and back. Six times.'

Her silent mouthing again. *The moon and back.* 'And they can fly *that* far?'

'Over a lifetime, yes.'

'That's impossible.'

But nothing's impossible. That's what I was secretly telling myself. Swifts prove as much: the migrations, the heights, the distances, the speeds, the ability to navigate from African jungle to the same tiny crack beneath the same fascia board in Harrogate. Miracles happen in nature every day, we just don't pay attention. Things disappear and things return. Cloudy mornings bring in sunny days. The world sometimes rights itself. Sensors pick up missing heartbeats.

And a few miles away, the half-hour of surveillance came to an end. The nurse shook her head with bemusement and began removing the belt. 'Nope, there's nothing to worry about,' she said. 'The baby's heart rate is perfectly normal. Strong and regular with just the usual variability. Spot on for twenty-eight weeks.' She even showed Rosie the data printout – a perfect pattern of peaks and troughs. A mountainous, panoramic heartscape. Understandably, Rosie quizzed her: *then why had it been fast? Why had it failed to register? Why these irregularities?*

'Hard to say. Could have been anything. I'll have them look into it. My guess would be a problem with the equipment at the surgery. Either that or human error.'

Then the baby kicked, hard, and, as though proving a point, it did so all evening, knocking and banging like the swifts outside our bedroom window.

Three weeks later, and they have been among us every day since, revelling in hot, blue, powdery light as June slips unnoticed into this high, burning July. They have taken over the skies and are screaming the place down. Their shrieks whip and whistle past open windows, through curtains and office blinds drawn at midday to try to keep rooms cool. Everything being done is being done to that sound. Power saws moan through planks. Scaffolders sing tunelessly to radios. Cars speed up and slow down. Seeds disperse. Lorries clang and grumble. Tractors spill new hay at traffic lights. People dress and eat and go to work and stumble home sweaty and tired. The bell of St John the Evangelist rings out over the edge-land. And above it all is a constant reminder of other rhythms at the margins of our lives. Even as I type these words I hear it through my attic window – *seeeeee-seeeee, seeeeee-seeeeee* – more females than males in that passing cluster, judging by their calls. It's easy to tell. The females take the higher register in their maniacal, whirling, rusty-wheel duets.

I see them all the time too. Occasionally, mid-sentence, I'll stop typing and look up at the square of sky in my attic roof. A swift will be framed there for – what, a second? No, it has to be less than that – half a second. Faster than a blink. A slide accidentally double-clicked through a projector carousel. Sometimes the bird will be buoyed high up on a thermal, flapping madly, finding its balance before powering out of shot quick as a jump jet; other times one will be skimming only inches above the tiles. If I could pause time, I'd see more than just a blur rocketing past, I'd see a sleek, dark, chocolate-brown body, almost oiled, like an otter's coat, and a tail that can either be forked and fish-like or folded into a tapering point. I'd see a faint creamy chin-strap and those kukri-shaped primaries: massive, blade-like, swept-back, a hindrance anywhere except in the air. I'd see an almost alien head perfectly adapted for open skies and, if I got closer still, an eccentric aviator's face: large, black binocular-eyes, like a pair of flying goggles with their own in-built anti-glare eyebrows. I'd see a tiny, soft beak – a moustache really – above a mouth that becomes more basking shark than bird when hunting, gaping to inhale insects in phenomenal numbers. What I wouldn't see, though, unless actually reaching through the glass and poking around its plumage, are legs. Swifts have the shortest stumps in the bird world, barely more than extensions of their bat-like, clinging feet, which retract completely when flying. Perfect for aerodynamics; useless for walking. Combined with those oversized wings, they mean that grounded swifts can struggle to get airborne again and, once fallen, will flop about, helplessly flailing for the air on the hard earth. People find them in this state sometimes, like a boat stuck on a sandbank, wings oaring away, the rower still going ten to the dozen.

Many years ago concerned neighbours called on my mother to help remove an injured 'something' making a racket above their living room, shuffling, scratching and screaming. They were surprised to learn it was a swift, recently fledged, that had free-fallen straight from its nest into the recessed flat roof above their bay window. Up a ladder, Mum found it imprisoned by the raised sides of the leaded lip, like a spider in a bathtub. She picked it up, cupping gentle hands around those air-filled bones, then held it above her head on the flat of her palm, lifting and lowering it gently. After a while, perhaps understanding the upstroke of open air on its face, it fell off her fingers forward and flew, pumping its wings and shooting up to join a screaming throng careening around a chimney. I always used to envy Mum that memory of holding and releasing a swift. So much so, in fact, that I once longed to find one stranded on the floor, just so I could experience the paradox of weightlessness and power, those long wings between my fingers, the heart racing in my palm, that worldly eye fixed on mine.

Not now, though. With my paternal instincts becoming keener each day, I just hope that the swifts have settled, mated and are incubating eggs. I thought I saw two coupling in mid-air above the ring road almost as soon as they had arrived on our street, but they were out of frame before my eyes had fully focused, lost behind the pitched roof of a betting shop. A good sign, though, is how annoyed the pigeons have become, how grumpy they seem about sharing their urban estates. On the opposite roofs they gather in clucky, sulky groups by the chimney stacks and gutters to mither: *These out-of-towners, bloody immigrants, coming here and taking our houses...mumble, mumble.* They were moaning so loudly the other day that I had to get up to close the window against their incessant cooing. As I did, I glimpsed a swift emerge from a crack just below them. It fell out, opened its seventeen-inch wingspan and masterfully looped-the-loop over the pigeons' heads. They all jumped and screamed, scattering with much shaking of ruffled feathers. I nearly applauded; I'm not even sure the swift didn't do it on purpose.

After mating, female swifts will lay a clutch of two or three small, white eggs, which the pair takes turns to sit on for around twenty days. Incubation is, by all accounts, a picture of domestic bliss. Jobs are shared: while one sits on the eggs, the other goes out hunting, joining the marauding gangs of a colony's non-breeding swifts racing in the streets and hunting insects. Unlike the younger birds, however, which climb ever skywards in the fading light to sleep in the open air, a breeding bird usually returns to its nest at nightfall and tucks up side by side with its mate. They announce this by screeching to one another – *I'm here! I'm back!* – late into the warm evenings, catching me unawares as I'm washing up with the window open or gathering clothes off the line in the backyard. Such gestures are wonderful, though. I find it heartening to hear their shrill homecomings.

The shock of Rosie's phone call that morning and the brief gravitational drag of loss it brought has taken its time to unknot from my insides. Although a false alarm, it was enough of a slippery footstep over a precipice for us both to pull each other back. A turning point. *Slow down. Hole up. Nest.* Rosie was overdue some holiday and she has taken it. Even though the glorious days continue, our time is spent home-bound, wrapped up with lists and chores, cleaning walls and carpets, finishing the decorating jobs I'd never got around to in January. And always sleeping early, side by side. The birthing books say the nesting instinct is a wholly natural stage; that, like birds, humans should 'listen to the body' and respond to our embedded biological imperatives. I think now that they may have a point. This morning as I waited for my coffee to brew, one hand resting on the cafetière plunger, Rosie surprised me, bouncing downstairs smiling, energy restored, eyes bright. She opened the newly cleaned doors to the yard, filling the kitchen with light and the screeching of swifts. 'I feel like a different person today,' she said. And standing there, washed by a beam of sun, I could see she was.

We walk down Bilton Lane two hours before sunset. The sky is still treacle-thick with warmth, a smothering density of bronze and blue filled with pollen and the smell of bracken and traffic vapour. There's no rush so we take the back roads, ducking under the low-hanging leaves of a lime sticky with aphid honeydew and thrumming with worker bees. Parked underneath, a van is sugar-coated, glazed like a doughnut and sprinkled with dust. I can't resist touching its bonnet. I used to do the same as a kid just to feel the skin of my fingertips take to its tacky lacquer. As we approach the edge-land, I feel another pulling sensation, this one between my eyes. Magnetic. Over the roofs and lampposts, past the tanning salons, takeaways and dingy-windowed newsagents there is a pylon-stapled join where sky ripples down and meets the limits of town.

These past weeks are the longest I have spent away from the edge-land and I can sense the change. I feel a pang of regret – jealousy, perhaps – that it hasn't waited for me, that I've missed something. As the space between us narrows, this becomes mixed with other feelings, a sense of returning, of absorption, of acceptance. Homecoming. I bore Rosie by relentlessly pointing out details: *Look at all these elderberry buds*; *I can't believe the size of the butterbur leaves!* But I can't. And I can't help saying it, either. There's too much to take in. I have to let a little out. Along the old railway the infinite greens of the bushes, hedges and trees have reached their limits and lie draped and meshed, sweating out the heady incense of hot herbage. Between them poke the delicate pink spikes of willowherb and the white cloudy clusters of hogweed flowers, crawling with longhorn beetles. Wasps throb on burdock leaves. Further on, in the meadow, there is the honey smell of clover; its flowers are dotted purple-red or white among the seed-topped grass and splats of dusty plantain. Knapweed frills the edges of overgrown paths. Tufted vetch too. And common vetch. I see the shaving-brush seed heads of hawkweed oxtongue. Woundwort. Stinging nettle. Evening primrose. Speedwell. Dunnocks burble deep in the hedges, wrens trill. Fences and wires are covered by bindweed and bracken. Beneath them, the deep black pools of shade. On the far, running horizon, great swathes of hay meadow have been cut. Idling tractors are primary-coloured blobs in yellow stubble. The air is sweet dust and dandelion seeds and the hot metal of gleaming pylons. A toad plods splayed-toed across the path into a dry stack of grass. Crushed underfoot, pineapple weed scents our every step. The earth sings with crickets. I point and point and point. And look. And hear. And smell. And all the time, in my head, there's a record playing, a particular phrase from *Tubular Bells*, Side 2, that runs from 4:21 to 5:21. This line selects itself sometimes on my internal iPod whenever view and atmosphere fall into place. There is something in the jarring 6/8 timing and the deep wistfulness of the piano part that sounds like the musical expression of joy and sadness combined. Its minor descent tumbles childlike across a landscape of guitar arpeggios that, at moments, echo exactly the chaotic, concurrent hums and squeaks that surround you as you walk here on a summer's evening. That feeling of being fully in the moment and fully aware that the moment won't last.

Wandering up from the lane, we are in the raggedy fields running up to the holloway among wheat that is tall, parched and paper-cut sharp. Nearby there is a little cave in the hedge formed by the ribs of an arthritic elder where six months back I hid, alone and cold, and watched the fox hunting over frozen fields. A month ago I sheltered there from the deluging rain. Now a strip of evening light through the foliage is filled with gnats bouncing around like atoms. And red blackberries. A billion stems of grass. Mass molecular engagements are occurring between the air and my skin, the light and my eyes, sound waves and my auditory nerves. The edge-land is so powerfully alive and glowing that I need to take a breath and stop myself from becoming tearful. Right now I can sense something bigger in the curvature of the horizon, the birdsong, the unearthly crawling of insects and the immeasurable flowers. Something exquisite, enriching, frightening, indifferent, immortal. And I realise it doesn't care whether I'm here or not.

Everybody loves nature, but I wonder if, deep down, what we really want is for nature to love us back. That's the impossibility of our obsession. It won't. It can't. I have come to know this patch of ground as well as anyone might know anywhere. I've seen through its eyes, immersed myself in its histories, dug through its layers and obsessed over its life forms. I've walked, crawled, curled up and slept on this earth. Even so, this evening I'm aware that there will always be something elusive and indescribable about this place. Something removed and separate; out of reach. For all the attention I've lavished on it, the edge-land cares no more for me than it does a speck of bacteria on the skin of a worm. That's one reason it remains so compelling, so haunting. The other is precisely the opposite: because all my time here has developed a curious feeling of shared history, some tangible, emotional intertwining. It's in my blood. After half an hour of being here, it's conquered me again. When people talk of 'knowing' or 'belonging' somewhere, this is what they mean. Familiarity comes with the overlaying of our experiences, memories and stories: *there's the stretch of river where the mayfly rose; that's the owls' nesting tree; these hedgerows were once the boundaries of enclosure.* We project all we are and all we know onto landscape. And, if we're open to it, the landscape projects back into us. Time spent in one place deepens this interaction, creating a melding and meshing that can feel a bit like love. In the drowsy light of the coming evening I not only see where I've walked before, but who I was when I walked there. What I was feeling; what I was thinking. And isn't this how we navigate this sphere? Creating fusions of human and place, attaching meaning and emotions, drawing cognitive maps that make sense of the realm beyond our comprehension? Our connection to the world is always two things at once: instinctive and augmented.

Rosie walks up behind me and links her fingers with mine. We follow the channel of a tractor's treads through the wheat to sit beneath the lone oak. The high, grassy curtain around its roots conceals us. The sun prickles our faces through its tickling stems. With my back against the tree and Rosie lying between my legs, we watch the day descend and share the bottles of warm beer I stashed in my rucksack. To the west, swallows flit, swoop and twitter across the surface of the wheat like flying fish. Higher still, in the upper air, the gentle whistle-screams of feeding swifts.

*

Later, cup of tea in hand, I'm locking up the front door
when a gang of them screeches past, returning to their nests
for the night. In social or territorial displays they've been
known to clock up 130 miles per hour, which sets me off
wondering, abstractly, if a swift has ever sprung a speed
camera. Unlikely, I'm sure, but then again I've seen them
crop up on TV shows before, sneaking into shot on
primetime soap operas and documentaries, even photo-
bombing frontline war correspondents. Naturalist Richard
Mabey wrote of being struck by an unavoidable allegory
about types of existence after seeing a swift flash past the
commentator during a report on the shelling of Beirut in
the 1980s. And, tragically, this isn't a rare occurrence –
neither the shelling nor communities, and swifts, getting
caught up in it. Recently I saw a journalist midway through
a bulletin from Libya being buzzed by that same telltale
shape, his voice momentarily drowned out by that
unmistakable cry: *seeeeee-seeeee, seeeeee-seeeeee*. I knew
instantly what Mabey was talking about – that juxtaposition
of gruesome, human-wrought horror and life-affirming
nature. The coincidental sound of the swift's plea seemed
to bring the madness of blowing up neighbourhoods filled
with civilians into even starker relief. As though we've
become so corrupted that even other species are petitioning
us to pause and take a look at ourselves.

There's a chance, of course, that the swifts on our street
are being purposefully recorded. Numbers have fallen by
as much as a third over the last decade and much work is
going into understanding this recent decline. Perhaps satellite tags bound around their barely-legs are drawing
digital flight maps on a scientist's computer screen somewhere. Maybe their returning tonight will trigger specially
rigged 'nest cams' into record mode. Such dedication has

helped us move on from the days when we believed swifts and swallows spent winter hibernating in the mud beneath ponds. Yet for all the insight our surveillance society can provide, the maps and methods with which swifts navigate their world are surely as nuanced, complex and hidden as our own. Much is hardwired in that little brain and body. You see innate urges beginning to manifest on those silent, black-and-white nest cams. At around a month old, swift chicks begin this curious routine of pushing themselves up onto their wing tips and holding up their bodies, as if doing press-ups or a core-strengthening Pilates move. The theory is that, before even experiencing flight, they are programmed to prepare for it because once they slip through that narrow aperture into the sky they won't return to the nest. In fact, they will be airborne for a minimum of two, three, perhaps even four years until reaching maturity, mating and first nesting themselves. Depending on how late in the summer they fledge, they may even leave for Africa straight away, fresh from the nest, untutored, possessing the supreme means to plunge into the unknown.

The precise *hows* of migration still hover at the edge of our understanding, although we can be fairly sure of certain things. Swifts interact instinctively with their environment, navigating by registering minute changes in light, airflow, moisture, pressure and magnetic forces. Direction, location and altitude are determined through the position of the sun, the stars and the earth's magnetic field, which, although invisible to us, wraps around the planet from pole to pole like a ball of wool. Birds are believed to be capable of detecting it via molecules in their eyes, beaks and neurons in their inner ears. Researchers think it may even appear as a cluster of stationary spots in their vision, like having an arrow permanently pointing north. As I click off the lights and climb the stairs to bed that night, it strikes me that the maps of the swifts' world must feature augmented layers too. Geographic adaptations that make sense to individual birds, repeat patterns consigned to memory, 'regional' maps that bear no resemblance to what's on the ground but which join together physical landmarks like mountain ranges and rivers with other waypoints and markers of the mind – the place where their nest was once blown to smithereens by an artillery shell, the spot where a flying termite swarm gathers yearly. Who knows? Perhaps when crossing some barren desert they feel the pull of those ever-churning concrete circles at Bilton sewage works. It's true that we're all only ever passing through this world, but part of being alive is that magical process of making it our own.

Here's the thing: in an era when there can seem to be a deficit of wonder, swifts are like the sky: once you start, you can't stop wondering about them. Frustratingly, though, their elusiveness and pace mean you rarely get more than a glimpse of what they are up to in a town. The edge-land, however, reveals a wider perspective. From under the lone oak tree there are vast uninterrupted evening skies in every direction; I can follow their sky-slicing insect frenzies for far longer, into dusk, into the dark. So, I start to return here too, conducting my own nighttime sorties from the nest again.

I like to think the flocks congregating above me are made up, at least in part, of the swifts from our street. I have a hunch they are taking a similar route down here to me, wheeling up out of the lines of terraces and crossing the ring road, flying down Bilton Lane, over the nest-free realms of the modern housing estates, to reach the insect-rich air above the scraps of woods, meadows and watercourses of the fringes. If we could compare them, I bet the jumble of lines on our internal maps would look pretty similar right now. In any case, they are feeding relentlessly, which I hope is a sign of chicks being born. I've tried to gather more direct evidence. The other day I stood listening for their mousy cheeps outside number 25 just below where a mating pair kept sweeping in and out in a tag-team sequence, but the street noise was too loud. After a few minutes of staring at the eaves, squinting and straining to hear, I realised a woman was looking down from a window. She had a genuine look of concern and a phone in her hand. I sloped off, over-acting that I'd had something in my eye, trying to look inconspicuous.

Sometimes on these balmy evenings, the swifts appear to be momentarily stacked in the hot, flat air like planes circling, waiting for runway clearance. But they are feeding still up there. Swifts will hunt at different altitudes for different prey: gnats, beetles, aphids, ladybirds, flies, moths, mosquitoes and tiny airborne spiders drifting on spun threads, designed to catch the wind. If it sounds like a varied diet, it's because it needs to be. Swifts are thought to disgorge up to forty meals a day to their young, collecting and consuming somewhere between 20,000 and 50,000 insects every twenty-four hours, packing them into dense, gobstopper-like balls that puff out their throats and cheeks. Yet more staggering statistics; yet more reasons you'd pray for this bird in your hand during a game of ornithological Top Trumps.

I see none of these details, though. All I see are the birds gathering then uncoiling like a knot of black rope, buzzing, zipping and tumbling through the sky. They tack and turn, firing off victory falsettos, trawling in such numbers that it confounds the eyes. As though stirred by their dynamism, the wheat shifts too, waving like a festival crowd one moment and then swirling back into stillness again – forming a surface of gold that rolls, ripples and slides towards town. Watching it, I think about something I haven't in years, the way oil paint moves under a brush. There were a few of us at school to whom the art department was a place to escape to. It was situated, as all art departments should be, in a garret above the sports hall, away from the main buildings and the business of education. It smelled of coffee, tobacco, heated Perspex, hot, cut wood and turpentine, and we chose to spend lunchtimes there, in amongst those smells. We were sixteen years old, learning how to use oils on board, squeezing it from metal tubes and chasing it into thick, wet blooms of life as the grey-stained clouds of north Leeds rolled across the department's windows. It was long before we knew of surface-focused postmodernism and that texture could be art itself. We sought *real life* in everything and spent hours manipulating blues, greens, browns and yellows into landscapes beyond the city; landscapes to be marked on their likeness to the real thing. Watching the patterns of swifts and breeze moving through the sky and wheat field, I see now that we were recording reality more accurately than we knew. Beneath the final layer of our finished works, now probably painted over, skipped or thick with dust somewhere, those fluid whirls mirrored exactly the way the landscape, the sky, all of us, exist in perpetual motion. If only I could go back and tell myself.

At 8:30 p.m. the next day, I walk out in boots, shorts and T-shirt, carrying only a notebook and pen, and after a hundred yards I am already too warm. The evening is ecstatic. The earth hums in octaves. Only the lane smells of damp, old caravans; arid, tinder-dry air presses everywhere else, even deep in the wood. Wave meets wave on the fields – wheat and insects merge and split. The birds are in their element: the swallows rattle and click like bats, cresting the crops and banking around the trees and pylons; the swifts sky-scrawl in the gassy blue. From the lone oak I can see for miles, off towards burnished moor, ridge and wood. Further, to where pale purple bleeds into sky. The sense of scale is difficult to absorb and harder still when I crane my neck up to watch the swifts, now so aloof and aloft. After an hour or so the sky behind them grows tufted with the wispy brushstrokes of cirrus; ice crystals zooming over town at 300 miles per hour or more and yet seeming so slow, almost stationary to my eyes. The drawing-down of the day blushes these white and grey filaments into reds and glowing gold, and the blue behind dims, like a curtain being drawn over the horizon. Another hour and the swifts are little more than misted dots against it all, gnat-sized in my binoculars, being recalled back to the heavens. They look like they're falling upwards, as though gravity has been reversed. Their sound fades skywards. Eventually they are indecipherable and, as if each was a drop of black dye passing through the cotton clouds, the cirrus darkens too.

And they will float up there all night in that cooling, loosening air, dozing on the wing. That's what I shake my head at as I walk home and, waiting to cross the ring road, glance up at the sky. A pair of flashing wing tips blip smoothly overhead through semi-blackness. *Somewhere between me and that plane, swifts are bedding down.* Between the sprawl and the EasyJets, they make nests in the air. Of all the impressive traits this bird possesses, surely this has to be the most extraordinary. When not nesting, whether migrating or over territory, swifts rest by climbing into high altitude – up to around 10,000 feet – and enter into a state called 'unihemispheric slow-wave sleep'. A neurotransmitter shuts off half their brain, keeping the other half functioning and alert to changes in wind and drift, ensuring the bird wakes up where it fell asleep. Or, if migrating, squarely on course. The left side shuts down first before swapping with the right, an alternation thought to be responsible for the bird's gentle swaying motion through the air as it rests, swinging like a baby rocked in a cradle.

Unbelievably, nearly a hundred years before this nocturnal rest-flight could be proven scientifically through satellites and tagging, an encounter had already occurred between human and swifts in this semi-dormant state. In his 1956 monograph, *Swifts in a Tower*, biologist David Lack recalls the experience of a French pilot in the First World War coasting with his engines off over enemy lines – an account that Lack notes was written-off for decades as 'absurd': 'As we came to about 10,000 feet, gliding in close spirals with a light wind against us, and with a full moon, we suddenly found ourselves among a strange flight of birds which seemed to be motionless, or at least showed no noticeable reaction. They were widely scattered and only a few yards below the aircraft, showing up against a white sea of clouds underneath... We were soon in the middle of the flock, in two instances birds were caught and on the following day I found one of them in the machine. It was an adult male swift.'

That faint and final *seeeeee-seeeee, seeeeee-seeeeee*. Yes, yes, I do see now, but the more you see, the more you want to know. That's the quandary. The more the maps seem to interlace and overlay. It all makes it harder to imagine them leaving. Then, one morning, you step through the front door and it's that time – the *turning* time. The moment the season creaks on its hinge and, by chance, you overhear. The start of a slowly spoken countdown; the intake of breath before summer's auctioneer yells, 'Going, going...gone.' Outside it appears someone has messed with the contrast and everything has become a little more light and shade. Leaves have darkened from the greens of new lawns and fresh limes to the hue of classic cars, a deeper, richer hunter-green, forest-green. Goodwood and old tapestries. It's hard to put your finger on it exactly. The heat and glare of the sun are just as stifling; the air bright and blissfully undiluted by icy northern winds; the leaves still weeks or months away from falling; summer holidays barely in full swing and yet, it's *there*, agitating the senses, a blur in the heat-hazed horizon, a shift in that swirling brushstroke, a tint to the palette.

In a normal year this is also the moment the swifts disappear. They are masters of 'ghosting' – the Hollywood-favoured technique for getting talked about by leaving a party early and without a fuss, slipping through the back door in the wee hours without so much as an air-kiss goodbye. The art of tactical absence. Fledglings tend to go first, followed by the adults and, before you know it, the whole party's on the wind-down. The streets are immediately lonelier. The traffic louder. The air dulled. Life is less interesting. And you stand there wondering why until, looking up, you notice empty sky.

I say 'normal year' because the chicks on our street will fledge late this summer. Rain has caused delays. It's almost August now; my guess is that, even if they're being fattened up fast, they might not begin migration until early September. And selfish as it sounds, I'm glad they're going to be here a bit longer. They've come to mean more to me this time around. I know their disappearance is a necessary one, to linger once the insects fade and the frosts come would be a death sentence, but the thought of it hurts nonetheless. The coming and going of swifts are markers in the cycle of our year. Their departure, whenever it comes, augurs an ending.

They've harvested the fields in the three days since I was last here. Now we sit against the lone oak again amid wheat stubble, dust and stones. The air smells yeasty, like a baker's shop first thing in the morning. The tree's roots cradle my ribs. With my arms around her, I, in turn, am cradling Rosie and her bump. I can feel her breathing, the rising and falling on my chest. I'm not sure why it would only just occur to me but I'm trying to straighten out in my mind the fact that soon we will always be three. It will never just be Rosie and me again. The long summer that has been our life since we met twelve years ago is coming to a close. And that's it, I suppose: the realisation that part of our lives is already over. The earth turns and the swifts move. You can't stop the ceaseless clockwork of the universe. Then, as I'm thinking this, the baby squirms beneath my fingers. I feel it like a knuckle twisting in my palm and shut my eyes to better concentrate. After a moment a foot, hand or elbow stretches the skin again. This time my heart lurches with it and I laugh. Rosie does too. Then I think, *How stupid of me*. How could I forget, out here of all places, in the fields of the edge-land, that every ending is also a beginning?

METAMORPHOSES

*Now I am ready to tell how bodies are changed
Into different bodies.*

Ovid, *Metamorphoses*

At midday a gust of wind blows briefly down the street. There's a change in the air. A pizza box skids a few yards, rattling its crusts, and flips up against the hubcap of a parked Nissan. For the last hour that greasy cardboard square has covered up a narrow crack between pavement and kerbstone. Now the sun pours down into the miniature chasm, touching the long, folded-back wings of an ant, a queen, as she clambers up its soil sides, using the withered leaf of a dandelion like a loft-ladder. Her antennae emerge, straightening and bouncing, reading air and concrete, followed by her glossy ink-drop head and body. Other ants bubble up behind – smaller males, similarly winged – each making short, darting runs, probing and returning to their fissure as though unsure, massing and conferring. But the queen has waited all summer for this particular conspiring of humidity, heat and light. She opens her wings and flies, flicking through the air quickly, unsteadily, touching wall, pizza box, tyre and tarmac, releasing chemicals to synchronise with neighbouring colonies and lure males into mating and swarming. Many are already airborne and, from out of the blur of daylight, one grabs her and they lock, tumbling along a windscreen wiper and onto a creosoted fence where a cat slumps. The breeze catches them again and they go with it, flying northwards over pavements and fences. A swift dips and ploughs a furrow beside them, mouth agape, swallowing twenty ants, but only glancing the queen and her mate with its wing tip, knocking them down through the tangled leaves of a goat willow and into the foliage that runs along the old railway.

A butterfly watches the ants drop into the patch of stinging nettles and writhe around on a leaf. With compound eyes consisting of thousands of honeycomb-shaped lenses, it registers the brief changes in light and leaf position, the oily roil of disrupted air and after-tremors in the nettle stems. But it remains still on the willow's trunk. With its hindwings folded to reveal flecked black and brown undersides, it is as good as invisible roosting on bark and yet it sees everything in its dome of vision: the scramble of the ants now unsexing, the male flailing backwards; the queen moving to the edge of the leaf, then flicking down to a morass of stone and earth beneath. It watches too as the queen contorts her body and, one by one, tears off her wings before tunnelling into the soil beneath an empty Coca-Cola can to birth a new colony.

The breeze lifts again, rustling the willow's leaves. Sun freckles its trunk. The butterfly senses warm pools of light on the bark above it and crawls towards them, crossing spiky patches of cambium where recently knife-gouged letters spell out the names *Joe* and *Lauren* separated by a heart. It opens its wings and sunbathes, soaking up the heat through its mole-like body, dark brown and covered with velvety hair. Its wings are the same shade, save for the stark orange-red bands and a few white dots, like dribbled paint over the tips. The willow shakes again. The sun dapples and disappears. Now the butterfly moves, churning the air and flitting sideways, landing on the same nettle leaf where the ants fell. Crawling to the tip it lays a single, tiny green egg and then rises, this time flapping westwards, dipping and climbing, keeping parallel with the old railway until coming to rest on a buddleia bursting with cones of purple flowers. Sense organs in the butterfly's feet immediately taste the sugar and it slips its proboscis into the open heart of the petals and siphons their nectar. Drumming its feet, it works its way up flower spikes, until a twinge in its abdomen sends it banking over the meadow towards another clump of nettles. It is halfway across when it detects a different taste in the air – *salt* – rare, but an essential mineral in its diet. Amid the meadow grass cut into fluffy lines between the old railway and the wood, a man lies dozing under the full glare of the sun. Beads of sweat bubble, trickle and pool on his brow. The butterfly circles, then softly shakes down to his forehead and unrolls its tongue.

I open my eyes. *A red admiral!* For ten or twenty seconds its silhouette blots out the zinc brightness of the sky as it absorbs the sodium chloride evaporating from my skin, then it floats off in that weirdly graceful-dashing way butterflies move, like burning paper. Maybe it's the heat – I've been lying here over an hour without hat or sun cream – but as I jump upright to follow its flight, the world swoons. The sun, a Maltese cross above the pylons, has scorched the asphalt and roofs, smearing the edge of town into a blur. Everything seems briefly skewed out of true, as if the butterfly slipped me something as it drank. Rubbing my eyes only makes it worse. Now every blink sends blobs of green, mirages of field and wood, flashing over the houses. *This is what happens* – I tell myself – *when you get up too fast. A momentary lack of blood in the brain. Blind spots. Disorientation.* I concentrate and try to bring it all into focus.

It is 1979 and Bilton is caught in a wave of change. The old railway has been decommissioned for years; its sleepers removed and track reduced to scattered shingle. The sidings are overgrown; the grand viaduct over the Nidd abandoned – a far cry from twenty years earlier when smart, stout 'Hunt'-class engines thundered daily across the gorge, whisking passengers between King's Cross and Edinburgh, each locomotive emblazoned with a leaping fox on its nameplate. Vanished too are almost all traces of the smaller, narrow-gauge line – The Barber Line – which hauled coal from the intersection of the railway and Bilton Lane ('Bilton Crossing') to the New Park Gas Works between 1907 and 1956. Now the only signs it ever existed are the immovable airshaft and a bricked-up tunnel peeping through the hydrangeas in a newly completed suburban garden. After years of vandalism and wet rot, the council has demolished the red-brick signal box that stood at the crossing point, its white wooden gates having fuelled the community bonfire on 5 November 1965. But such details are minutiae. They pale into insignificance when compared to the scale of the transformations happening to the south.

After decades of building momentum, the town's encroachment over Bilton's open fields and ramshackle cottages has reached fever pitch. Construction is booming. Tightly packed micro-suburbs and rural-sounding housing estates – Sandhill, Meadowcroft, Woodfield, Coppice, Hill Top – explode like flower heads from the tendrils of Edwardian and Victorian roads that once demarcated Harrogate's urban limits. The farmhouses, meadows and cornfields that had conspired to form a landscape here since the Enclosure Acts have all been sold; most have been steamrollered and built over. The remaining few await their turn. Divided and sub-divided again, the estates fold into confusing patterns of beige, grey and red, hotchpotches of style and design that spill across the gently undulating slopes. Link roads, recreation grounds, community halls and new Lego-block schools cut across the old views of distant hills. The sprawl has spread so rapidly since the Second World War that it has begun to eat its own tail. Ageing veterans of Dunkirk and El Alamein who bought the first pebble-dashed semis after demob grumble to each other about mullet-haired plasterers with their noisy transistors singing along to Gloria Gaynor's 'I Will Survive' over their back fences. It seems that every time they look a new avenue of concrete-rendered bungalows has sprung up; some man in flared jeans is hauling a tea chest through a front door, a stack of Sabbath LPs under his arm. Even where the boggy fields meet the old railway's sidings, a building site has been delineated with fencing. Diggers are tearing up the last of the hawthorn hedges and drystone walls, preparing the earth for what will one day become the smart cul-de-sacs and driveways of the 'Poets Estate'. There's an irony to this, but one perhaps lost on the glum-faced digger driver hacking away great sections of the tawny clay: the line of the railway was originally drawn far out here specifically to ensure it kept its distance from Harrogate. Now the town grabs at it like a lifeline. Once the houses of Keats Walk, Tennyson Avenue, Shelley Court and Coleridge Drive are finished, it will describe its northern boundary.

A policeman picks his way around the cluster of bare-chested builders sitting smoking rollies on a stack of bricks. The copper is new to his job – it's clear from the pressed uniform and the mirror-shiny boots. The valiant attempt at a moustache on his upper lip can't disguise his greenness. One of the builders wolf-whistles, sending his mates into fits of giggles; spurred on, another one enquires loudly, 'Who ordered the stripper?' The copper says nothing but grits his teeth: this is not how he envisaged life in The Force. This is not the kind of action he dreamed of when watching *The Sweeney* with his girlfriend over Saturday-night egg and chips. But he slops through the mud regardless, reminding himself that solving crime is solving crime, whether heroically apprehending a murderer in gangland London or investigating reports of a stolen car dumped on the outskirts of Harrogate. Squeezing through two sections of steel-mesh fencing, he athletically scales the bank in four leaps. From the railway, he scans the land to the north, frowns and pulls a map from his buttoned-down pocket. He's never been here before but judging by the symbols, even the cartographers at Ordnance Survey have struggled to keep up with the pace of change. The remnants of track under his feet are pretty much the sole point of correlation. Where the map details a meadow and, beyond it, a plump wooded gorge, reality reveals a swathe of sad mud torn up by motorcycle scramblers, dumped with mounds of rubbish. Brick pallets and paint pots have been hurled up here from the building site; chipboard and packaging stacked against the gorse shrubs, soaked with petrol and set alight. A noxious stench drifts from blue plastic drums kicked onto their sides, and the pools of liquid inside swirl with the rainbow sheen of chemicals. Oxidised tin cans and the glassy detritus of broken bottles are scattered through the flowering knot-grass. The policeman strides out, keeping half an eye on where he treads, walking across a patch of earth old as England, now scorched, weedy and strewn with torn-out flaps of kitchen linoleum, and fag packets.

Town-bruised, town-battered, the birth of the edge-land has been a messy one. Quick, but bloody and injurious. Its woods, once the sylvan heart of Bilton Park and the Forest of Knaresborough, are thinned and sickly, the edges deformed, hacked down and fire-scarred. The copper detects a whiff of fresh smoke and smouldering plastic as he steps over the dusty, tattered semi-skeleton of a crow. The gully marked 'Bilton Beck' on his map is so soupy with litter it is more scrapyard than watercourse. The wreck of a motorbike juts up from the water like some installation from the Indica Gallery. The grey shell of a burnt-out car sits halfway down the bank, engine exploding with brambles and sycamore saplings. There's no point scrambling down for a closer look; it's too old to be the vehicle he's been sent to investigate.

Half an hour later, in the meadow further west, closer to the viaduct, he spies a more likely suspect. The pea-green Austin Maxi has its doors flung open and is wrapped around the trunk of a Scots pine, half-hidden in a bed of stinging nettles. Notebook in one hand, the policeman reaches in with the other and removes the keys still dangling from the ignition. Then he carefully records every detail: registration, condition, location. He checks the boot and scrutinises the steering wheel and gearshift for fingerprints, then bags an eight-track cassette of Pink Floyd's *The Wall* stuck in the stereo, and a cigarette he finds tucked under the carpet of the passenger foot-well. Evidence. It is a thorough inspection, yet one that fails to record the casualty at the scene. There is a body. Crushed between the Maxi's radiator grille and the tree, a partially mangled creature is burial-wrapped, shrouded by its own wings – beautiful brown wings with orange streaks and white spots. It will never be discovered. Long before the policeman can organise a salvage team to recover the car, the errant hands that hotwired it will be back to finish the job, dousing its seats with lighter fluid and tossing in a screwed-up torch of flaming newspaper. Seconds later, in the fireball that engulfs it, the last red admiral of Bilton will be instantly cremated, lit by the fiery light of exploding modernity. Butterfly, sacrifice, metaphor.

I blink. I am sitting in the same meadow, unpoisoned now, healthily unkempt, brimming with life and insect noise. The heat has eased. The sun has been taken down a peg or two, calmed by the faint procession of dirty-hemmed clouds drifting north. Around me the landscape settles back into familiar forms of the present: mounds of mown meadow-grass, the brown, seed-clustered docks, the tall thistles, the silvery cotton-wool thistledown, vetches, plantains, nettles, trefoils, lesser and rosebay willowherbs, pink balsams, ripening elderberries and bouquets of Bird's-Eye-custard-coloured ragwort. The air glitters with trails of flying ants being chased by black-headed gulls. Bindweed turns a dead hogweed stem into a shaggy Christmas tree; a white bellflower sits like a star at its top. Insects explore the forest of hairs on my arm. Red soldier beetles dotted with Day-Glo pollen mate on a cow parsley. Abundant, wild and bright with August light, it is hard to imagine that this shabby utopia could ever have been the tormented earth of that previous scene.

I pinch between my eyes and rub the back of my neck. *Up now.* I'm thirsty and need to stretch my legs. I head over the lines of grass to the old railway, flicking the flies away from my ears as I walk. Goldfinches, goldcrests and linnets loose sparkling phrases from the hollies, birches, hazels and willows that narrow this stretch of the old railway into a leafy carwash. It's all so absorbing, the leaves, the wild melodies and improvised lines. Because they were here when I first discovered this place, I've always (unthinkingly) imagined these shrubby trees as permanent fixtures of the edge-land. Now, as I move between their branches, I realise not one of them is even as old as me. This is the deception, the lie of the land. The feelings of agelessness, the firmness of the earth underfoot, the infiniteness of the moment – all can draw in and dazzle the eye into believing in the ongoing fixity of the present. But nothing ever stays the same. Still, it may be, but still moving. Blink and you miss it.

The heart-shaped nettle leaf might have been cut with pinking shears. The edges are a perfect sawtooth pattern blemished only by the pinhead-sized green orb glued to one side. A week of warmth has passed since the red admiral curled its abdomen and laid it there. The egg has darkened; a tiny bulbous-headed caterpillar – a larva – squirms within. Daybreak is already midway through its performance over the edge-land. Faint shades of blue and red are a diaphanous backcloth; the gloomed, brooding trees are shot through with early sun. A chaffinch jigs between the willow's branches then entwines its song into the woven notes of the dawn chorus being threaded above. A moment later, a hole appears at the top of the egg, widening and widening as the larva pushes headfirst through the waxy protective layer to split open the outer chorion.

It bobs around for a moment, absorbing its new environment, processing the changed intensity in light, the shape of the air. Then, like a jack-in-the-box, half-in, half-out, it works free its segmented form and slips from the egg completely. It straightens as it crawls along the leaf. With its thin cream body and black head it resembles a minuscule struck match, blown out before the flame could sully the wood. At its scale the nettle's surface is covered in transparent needles, each filled with concoctions of formic acid, histamine, acetylcholine and serotonin designed to blister and wound. But the stinging nettle is the red admiral's host plant and the larva moves knowingly and untouched to tear at the tender sections of green in-between its barbs. It feeds ravenously, trying to sate an incessant hunger that is designed not to sustain its life but to change it.

The light swells. The tree-gloom sinks into the soil and the dew dries. The edge-land is in full voice now and the town modulates with accelerating and decelerating traffic. A blue tit flies into the lower branches of the willow and spies the larva feeding in the nettles below, hopping down six inches to get a better look. Then it loses interest. In its eyeline, attached to a willow stem, is a more nutritious prize. It tilts its head, darts up, and wheels back to its nest with the fleshy green body of a hawk-moth caterpillar curling around its beak. The bird's flight shadow whips across the nettle patch, crossing the already swelling larva, stirring it from feasting. Instinctively it moves, rippling to the side of its leaf, and begins to dispense a thread of strong silk from a spinneret in its lower lip. Working quickly, it weaves the leaf's two edges together, drawing its halves upright and inward on themselves, sewing up the join from the inside. This tent will be the first of many it builds in the coming weeks for concealment and shelter. A folded, sewn, edible nest; a place to hide, moult and alter.

Instar – from the Latin 'form' or 'likeness' – is the word for the physical transitions the larva undergoes as it grows. Because its skeleton is on the outside of its body, each instar has a pre-set capacity, an unyieldingness that must be overcome in the name of progress. Even now as it eats away relentlessly at the furthest edge of its leaf tent, the larva's exoskeleton is tightening, its birth body filling and nearing its limit. Processes are occurring under the surface. Enzymes are being released. Molecules are being modified. Subcutaneous skin cells are already detaching from the outermost layer and beginning to be reabsorbed. Reorganised, recycled, this matter will form a larger exoskeleton manufactured within, soft and folded up like a parachute. Soon the larva's previous shape will be nothing but a thin sheen to be shrugged off and shed as the new incarnation ruptures it, expands and hardens.

Because of their fine-scale integration with landscape, their dependency on particular host plants and their recognisable forms, butterflies are what environmentalists call a 'key indicator species': a reflector of the health of a wider area; a being through which the land might be read. By zooming in and studying them, we can zoom out and take the pulse of a place.

It was a full year before anyone noticed the absence of the red admiral in Bilton. Then the following summer, 1980, it started with a casual enquiry: Bill Varley – a resident of Sandhill Drive – remarked to a neighbour that he was yet to see a single one that year. Tortoiseshells, peacocks, orange-tips, the ubiquitous cabbage whites all over his brassicas, but no red admiral. *What about you, David?* His neighbour shook his head. 'Can't remember the last time, come to think of it,' he said, crouching to uproot some groundsel from his petunias.

Intrigued, Varley abandoned gardening for the day and began asking around. Pottering from neighbour to neighbour, he quickly established that not a single person on the street had clapped eyes on one. It was the same story on the next street, and the two after that. It soon preoccupied him enough that he bought a notepad and started to keep records. Daily diaries. His search grew into a kind of obsession, the sort that often sprouts into the mind of a recent retiree who suddenly finds himself set adrift and devoid of the rigours and responsibilities of the nine-to-five. He started enquiring further afield – his neighbours, at the Post Office, the bar at Bilton Cricket Club and the regular coffee mornings and jumble sales his wife dragged him along to at St John The Evangelist's. Eventually he bought some advertising space in the *Harrogate Advertiser*, titling his entry for 8 July 1980 'The Strange Disappearance of the Red Admiral'. *Residents of Bilton!* – it began – *Have You Seen The Red Admiral Butterfly This Year? Image below. If So, Please Contact Mr B. Varley. Harrogate 601436.* He received one response. It was from a lady living in Keats Walk, but they soon established a case of mistaken identity. The butterfly she had freed from her bathroom the day before had undoubtedly been a tortoiseshell.

After swotting up on the subject (cricket was his primary focus outside insurance for forty years), Varley suspected the red admiral's decline was probably down to the conversion of the land into housing. Properties like his own well-kept bungalow had completely covered the proliferation of flower-rich pasture and hay meadow that had existed in this spot thirty years before. Neat, manicured gardens patrolled by the allied forces of lawnmower and weedkiller ensured nettles had largely become a plant of the past. He also suspected – quite rightly – that the butterfly had been squeezed from the other direction too. Throughout the 1960s and '70s, the farms to the north had turned over all available ground to intensive agricultural methods. Hawthorn and blackthorn hedges had been treated, grubbed up and ploughed in to clear the way for new irrigation systems. Areas where stinging nettles, dogwood and nectar-rich wildflowers grew were lost as one field bled into the next and were sprayed with organochlorides to produce high-yield, monoculture crops. Varley reasoned that, if thus pressed on two fronts, the red admiral's last hope of refuge had to be in the space in-between.

He visited the edge-land on eight occasions that late summer before conceding defeat. Each immersion returned him home more dejected than the last. Sandwiches untouched. Notebook filled with scribblings. Perspiration thick and stinking. 'It's appalling down there,' he told his wife Vanessa, heeling off his wellies at the back door. 'Frightful.' As he stewed in the bath, she gathered up his clothes for the machine and listened at the door. His behaviour was not exactly concerning her, but it was different. A change. She'd worried about this happening when he retired. Gladys, a friend at church, had had the same with her Ian. 'Humour him,' she'd advised. 'Let him know you're interested. Otherwise you'll lose him. He'll secrete himself away and then one day, you'll wake up and you won't recognise him any more.'

And so she did. It was there on the dinner table when he came down for his beef stew, the flyer she'd picked up after having her hair set in the salon up the road. Typed across its top was, THE BILTON CONSERVATION GROUP.

'They're looking for volunteers,' she said. 'Thought you might be interested.'

'I certainly am,' he said, reading it twice. 'Thank you, dear.'

A peck on the cheek too, over the washing-up. *Good old Gladys.*

Contrary to what the name might suggest, there was no real conservation plan at first. It was more a feeling among a growing group of Bilton's more proactive (and largely senior) residents that something, *anything*, should be done to address the decline. The marginal land lying broken and burned beyond the back fences was intolerably damaged and increasingly dangerous; it was an eyesore for locals and a headache for the police. But as Varley found out at the regular Wednesday-night meetings, the first hurdle was trying to understand who had responsibility for this liminal space. The speed of development had shifted and blurred the lines of ownership. The edge-land was as undecided and unacknowledged as it was unloved, presenting both a physical and a philosophical challenge to the group. *Where was it? What was it? Who owned it?* Technically the council had claim over much of the neglected area, but it showed little interest in preventing its ongoing destruction. It did, however, do the paperwork. Sections of land were prospected and reclassified, inadvertently recalling what was buried beneath. Descriptors such as 'common land' appeared on the new maps where commons had existed before, like old wounds leaching blood through new linen. One patch was even designated 'waste', reverting it a thousand years to when the Domesday Book had written off all Bilton with the same word.

There were other aspects of this new edge-land for the group to contend with: the scruffy farm fields, the lanes, the pylons and the viaduct. Perhaps most significant was the wood that contained the wide sweep of the River Nidd and its litter-clogged tributary, Bilton Beck. Varley offered to do the research and discovered that the wood had been sold off along with the rest of Bilton Hall when the estate was broken up in the years after the Great War, its sole male heir having been killed in 1917. Legally it was in the care of private hands. But, as Varley related to the meeting that week: 'Clearly they are hands unwilling or unable to take on the restoration work it so desperately requires.'

Exasperated by the process, the locals eventually staged an intervention. Over the long summer of 1982, the Bilton Conservation Group removed more than forty tonnes of rubbish from the edge-land's meadows, woods and waters. Then its volunteers tackled its numerous burnt-out cars and motorbikes. The unlikely-sounding 'Harrogate Sub Aqua Club' was roped in to attach underwater cables to three saloons dumped in the Nidd. After some gentle badgering the army provided the apprentices and kit to haul them up the gorge's steep sides. Vanessa had come down on that afternoon unannounced, turning up with Gladys and bringing welcome Thermoses of tea, mugs and bags of scones *for the troops*. Watching his wife handing out the cakes on paper plates, all red-faced in the syrupy air, flustered by compliments, her husband had beamed with pride.

'Any red admirals yet?' she'd asked him, handing over a scone in a napkin then straightening the tie beneath his green tank top with a pat.

'Oh, I think we have to wait a while yet,' he said. 'But I'm hopeful now.'

Then, suddenly, catching himself by surprise, he kissed her. Equally surprised, she blushed and smiled. The Red Admiral (*Vanessa atalanta*). A derivation from the eighteenth-century common name 'Red Admirable'. *That's you, my dear*, Varley thought. *My admirable Vanessa.*

The clearing of edge-land was completed around the middle of August, whereupon the priority became ensuring there could be no future fly-tipping or stolen vehicles dumped and firebombed. 'The cancer is gone,' said Varley to the group, 'but we must now turn our attention to preventing its return.' And all harrumphed in agreement. So, by letter and in person, the group lobbied for adequate barricades that could stop motor traffic invading along the old railway, while permitting the open-foot access to those who wanted it. Impressed with the surprising strength of local conviction, not to mention the dramatic and cost-free improvement to its lands, the council duly obliged, installing concrete-footed fencing and a padlocked gate at the junction where the old railway intersects Bilton Lane. The earth stirred. Another inadvertent echo: a gated crossing point constructed where one had stood before.

As Varley leaned on it one evening, he took in the open waste ground beyond. Green shoots of dandelion, fat hen, shepherd's purse and stinging nettle were beginning to scab over the mud and burned earth. Even in the vanishing days of summer the new was already emerging through the old. An invisible membrane had formed itself beyond the fence. Although he would never admit to anyone, each time he passed through it he felt he was somehow experiencing a transformation.

Precisely how long it took for the edge-land to assume

its current form is difficult to determine, but the healing happened relatively 'naturally'. The Bilton Conservation Group possessed neither the funding nor the experience to launch the kinds of conservation or preservation strategies employed with landscape today. There was no appetite to locate the ground in some nostalgic frame of yesteryear, no rigid profiling of flora and fauna allowing only that which had thrived here when it was a royal hunting park. Nor was there a restoration of its rubbled flax mill into a visitor attraction. This was not prime countryside by anyone's definition but still a sweep of scrappy land where, even in its most remote spots, the shadow of the sprawl breathed down its neck. No, what happened was largely a letting-go, an end to the human interference or management, aside from the few still-worked farm fields and the council's yearly cutting of its hay meadows. And, left to its own devices, the land reassembled itself, repairing its DNA and becoming wild, lush, untidy and beguiling. Out of the ashes, its trees shooted and spread; the wood thickened and bloomed. Bacterias bloomed. Air and insect propagated its pollens. Brambles, bindweeds and cleavers crept. Blackthorns and hawthorns budded. Thickets tangled. Grasses and wildflowers seeded and carpeted. Native and non-native plants abounded and contested. Nature's rebellious children found their own balance, reclaiming wood and meadow and railway and river, the colonisers, the opportunists and the crawlers evolved into ecosystems unhindered, luring back life in its many forms.

The Bilton Conservation Group continued to keep a close eye on things. Volunteers like Varley walked the old tracks and attended to the edge-land's details through field notes. The group even drew from these to produce regular typewritten and hand-illustrated newsletters for distribution among its members. Today these read like short dispatches from inside some newly forming state, which in many ways, of course, they were. Each an astute phenological document, far more than the mere list of weather conditions and species it appears at first glance. In context, these acts of recording and veneration captured the development of this unplanned, unburdened and largely unnoticed realm as its shape emerged.

Sadly most of these reports were lost in the intervening decades. As the group's original members aged, downsized, moved into nursing homes or died, files and boxes containing these treasures became rubbish clogging up the corners of spidery lofts. Detritus to be cleared on Sundays by mourning families. The earliest surviving newsletter (No. 17) dates from winter 1985/6. Even so, being written only three years after the clean-up, it testifies to the speed and spontaneity with which nature had reclaimed the ground:

A biting east wind brought the temperature down to -1 °C and snow flurries encouraged us all to trudge briskly through 4" of soft snow. Six traps produced two sleeping Field Voles (Microtus agrestis) *a hyperactive Common Shrew* (Sorbex araneus), *a Wood mouse* (Apodemus sylvaticus) *and, surprisingly, a stray house mouse* (Mus domesticus). *Fox and rabbit tracks in the fields near the rim of the Gorge. A gibbet of Grey Squirrels and Woodpigeons reminded us of the endless battle the farmer has to control vermin. Time and again we were to catch Bank Voles* (Clethrionomys glareolous) *– a larger cousin of the more common Field Vole distinguished by its longer body (8") and tail (1"). Few birds were on the wing apart from Robin, Blue Tits, and Finches skulking in the thickest scrub, and an occasional Crow. With chattering teeth we* *examined the last traps of the woodland fringe before descending to the riverside. Here the steep slopes were shrouded with dormant hazel, beneath which our last captives – Wood Mice, Bank Voles and Common Shrew – blinked in the daylight before bounding away from the sharp eyes of predators such as Kestrel or Fox. The river was grey, cold and in spate. Alders trailed icicles in the current. Someone remarked, 'Who'd be a trout in winter?' and no one braved the flood to check for crayfish hibernating under the banks. Ducks and wildfowl have become less common and a wet, sandy spit provided evidence of Mink footprints, which might explain why. The Mink may fill that ecological gap left by The Otter, which is thought to have become extinct in The [Nidd] Gorge sometime in the 1950s. Escaped Mink wreak havoc amongst the native waterfowl and mammals. More encouraging though were the prints of Roe Deer, Rabbit, Fox, Stoat, and even a Common Hare, which we were fortunate enough to encounter before the walk was over.*

If not an exact description of the edge-land I found on New Year's Eve, it is a damn close likeness. An instar.

These long days. These late-summer days, drowsy, immense and golden. It's Tuesday. I walk up the lane at lunchtime troubled by the thought that I may have been lax in my own recordings of this place. The microscopic details of the here and now seem to possess an inexpressible value that I'm worried I've overlooked. I wish I'd kept more rigorous data. More snapshots. The changes in a single leaf in a single location from day to day. The biodegradation of a discarded fag butt on the stone track. The minute-by-minute movement of a single bird through the wood. Maybe these are the things of true importance.

But where do you begin? At the entrance to the holloway the hedgerow is dense and wild and inscrutable; it is almost impossible to make out the rotted stumps of fence, the animal runs, the tumbledown wall stones concealed within. I can smell wheat and death and sweetness. Meadowsweet, perhaps – the scent of the back of old cupboards and ground almonds. There's a scorched, fumy-blue sky crossed with wires. Starlings whistle. Dog roses (white). Foxgloves (pink). Shrivelled sloes. Hawthorns heavy with berries. A skyline of soft hills. Plane contrails crossing the pylons. A million things. The time is 1:54 p.m. I kneel and try to imagine a rough metre square in this section of the hedge and begin to document what would fall within its borders. My eyes get lost in the spectacular variety and I fill three pages in scribbled shorthand. When you take down the world in this way, it feels as much about holding onto something as understanding it; in some small sense you are always trying to save a moment. An indirect preservation. But by the time you finish each word or sentence, that world has already gone. Then a thought comes: perhaps in the future someone, *something*, will scour back through such records, read them for clues, try to decode the point where it all went wrong. They will celebrate nerdy field notes precisely for their banal lists and lament where description is trimmed and tailored in favour of style. By trawling the records of such microcosms, they may unlock the mysteries of incomprehensible macroscopic human behaviour. The interzones will serve as a legend to the wider map, testifying to our collective denial, our ruinations, restorations, contradictions and wilful amnesia towards our environment. Unearthed sometime in 3000 AD, a few pages about a single patch of edge-land might prove as vital and telling as all the scientific data on melting ice sheets and rising sea levels.

Down the holloway, just where the banks rise and the path feels suddenly cocooned by the earth and silver birch, a tawny pool of sun. I stand there with the place to myself and listen. The sound of the weir and the rusty call of a woodpecker. The air smells of coming rain. I descend the track towards the river, going deeper into the wood, passing the exposed gunnels, the mined pits and the quarried stone. Here, where the mill once stood, drift the ghosts of industries gone back to green. The ash circle of Sir Hare's fire is lost now under a mat of ground ivy. A place where I watched him wake and wash in the river, all bony and long-limbed, is a mass of flowering honeysuckle. Dog's mercury furs the rutted millers' paths as the ground swallows the last of their pebbles. Foliage prints new, exotic patterns on the earth. Bursting balsam pods fling their seeds through the undergrowth and into the folds of my clothes. Crunched between teeth, their tiny black or cream dots taste of water chestnuts. Across the river, the punky shock of a skunk cabbage sticks out among a bed of nettles. Its huge yellow flower looks both prehistoric and from a time beyond our own. Like the wildly re-grown meadows and the re-colonised old railway, these things point not only to a past, but a future. In them you can see a world without us when nature calmly repossesses whatever we have left.

I'm watching a heron stand statuesque in the shallows of the Nidd when there is a crash behind me. I spin and see what looks like a clod of soil hurtling through the canopy, separating into two. As I scramble up the bank for a closer look, the larger section moves from its position in a beech – a kestrel. It flaps away through a rift in the leaves back into open air. The second projectile smashed into a large holly at the top of a bank and looking up into its heart I see a blue tit, fluffed and shaken but not injured. It looks down at me from its high branch. *Did you see that?* written all over its face. *Bloody hell!*

'You'll be all right,' I tell it and off it shoots, dipping low and twittering.

The rain will be here soon so I cut across the meadow to join up with the old railway. A man looms out of the trees. He's old and grey-haired, but youthful in step and with a trendy T-shirt – a Union Jack made up of ripped and sewn garment fragments. He leads a fat, butter-coloured Labrador, which limps along behind. As they draw closer I see the man is holding a takeaway coffee cup and wonder where he could have got it. Then I see clamped between his side and his arm are three cans of fizzy pop, a crisp packet and a half-filled bottle of mineral water. He slows to talk with a shake of his head.

'I can't understand how some people think this is accept-able.' His accent is broad – South Yorkshire. He thrusts his litter collection at me. 'Thoughtless bastards, the lot of them. I've just picked this up from the verges. Can you believe it? And don't even get me started on the dog shit.'

I don't, but he's off again anyway.

'Every day I take at least two bags home. And that's not including my own.'

'That's terrible,' I reply, hoping that by 'my own' he's referring to his dog.

He shrugs. 'Yeah, well, I can't stand seeing the stuff. It ruins the place. Have you seen much today?'

I tell him about the kestrel and the blue tit.

'I meant litter.'

'Oh, no. Not really. A cigarette butt …'

'Yeah, well. There should be litterbins and dog bins all

the way along this stretch. And a seat or two wouldn't go amiss, either. But you wait. It's going to be better when they put the cycleway in.'

I had half switched off but his words are like a jolt. Fingers stuck in my sides. 'The what?'

'The cycleway. It's going right along here.' He sweeps his arm side to side to denote the length of the old railway. 'Didn't you know? It's been in the papers.'

No, I didn't know. I follow his arm as he swings it again, pointing left then right, towards where the track runs into the wood and the gorge, where the reinforced metal shutters block off the viaduct and prickle with spikes.

'Oh, they're going through all that,' he says, reading my mind. 'Pulling it all down. Sorting the viaduct out and extending the cycle track all the way to Ripley.'

I must look shocked because he frowns and attempts to gee me up. 'It's a *good* thing,' he says. 'They're doing the place up.'

'Who is?'

'Council. And they're starting soon. Cutting all this back for a kick-off.'

I have a thousand questions, but can't get any of them out. I can smell meadowsweet again. And hear the linnets.

'It's a *good* thing,' he says again. 'It's so that all them kiddies and families can come down and ride here, y'know? So they can get outside and into nature. It's a *good* thing.'

The sky is blurring and shading grey.

Bored with me, the man wanders off. 'Come on, Jess. It's gonna rain.'

But there's no need for the future tense. Specks are falling heavily on my coat. They thicken quickly into a downpour, worrying the leaves and shushing the fretting nettles.

*

Each rapid alteration in its size, colour and appearance has been profound. Moving from leaf to leaf, constructing and devouring, the caterpillar has changed entirely: the little body that emerged from its egg has swelled and thickened inside a succession of spun, silken nettle-wombs. Now, in a lull in the weather, it slips from its leaf tent in a final larval shape: a hefty, armoured, bullet-headed form that muscles over the nettle, weighing down the wet foliage. It is the same black as a male adder and trimmed likewise with zigzag yellow along its flanks. Each segment of its body bristles with defensive hairs, repellant hairs, thick, sharp tufts like gorse thorn. A chaffinch in the willow sees it, but doesn't move; of all birds, only a cuckoo would dare swoop and take it like this, but there's not been a cuckoo here for thirty years.

The air brightens, pearlescing the clinging damp that has dogged the old railway for days. A linnet sings hurriedly, as though thrust into a spotlight. The caterpillar continues to traverse its mutilated plant, winding down and down the central stem to reach the lower leaves of this sodden, hot clump. These are plumper leaves, but the caterpillar has no interest in eating them. Instead, it once again rolls up the edge of one and draws it together with another, binding them with wraiths of silk to form a cave of green. The sky darkens and a dog brushes past, shaking the stems. The rain begins again, drumming with increasing ferocity and running down the outside of the leaf cave to flood the shining mud. The caterpillar, a full three and half centimetres long, scales the sides to sew a small button of silk – a cremaster – between the leaf's veins, like a spider's web in the rafters of a barn. Into this it weaves its hooked hind feet and then slowly, as though testing its tensile strength, extends itself down, head first, until it comes to a twitching rest in a perfect 'J' shape.

Light then dark. A day passes. The light returns with dawn rain. A brief interlude of sun, then rain again. A ripple slips down the length of the caterpillar. And another. It becomes a slow, subcutaneous rhythm. Each tremor stirs the dark skin and reveals it to be just another thin membrane. Then a split occurs, silently, behind the head. The ripples continue like shock waves, pushing out the newer, tougher incarnation of a blunt-headed, immobile pupa – a chrysalis. Every contraction further releases it while rolling up the caterpillar's old form until it becomes nothing more than a crumpled garment of segments, legs and head gathered at its feet. The chrysalis gyrates, twists, and kicks it off, as if stepping out of a bathrobe. Then it hangs there unmoving; a strange, ergonomic-looking shell that would fit well in the grip of a tiny hand, with its smooth plates for the palm and ridges for fingers. In days to come, it will grow nippled and sow-bellied as it dries and turns the colour of an old leaf. Inside, disintegration has begun. Digestive enzymes will liquefy all but a few hard structures as the caterpillar's cells are recycled for a final purpose. Outside the rain keeps time in the leaves – *tick-tock-tick-tock*. A roll of thunder resolves into a diesel engine slowing to a stop. A Ford Transit flatbed, the council's tree-cutting truck, parks up at the bottom of Bilton Lane. Its occupants sit and wait for a break in the weather to begin their work.

'Everything up to the ash. Stop at the ash,' bellows the older man. If he's not in charge, he certainly gives the impression of it. He is talking to a younger lad who looks a little confused. There are two others further on, dressed identically: high-vis tabards and white hard hats. Ear defenders fixed over the top and toughened Perspex glasses.

They have the same squaddie tan – the arms and neck. They know what they're doing, though. Their chainsaws drop from high whines to resistant growls as each bites into the bark of bigger trees down the track. Two cuts either side, one higher than the other, and they step back. A silver birch slumps backwards with a crash.

'The ash?' shouts the young man.

His mentor coughs and wipes some dust from around his mouth. He walks over. 'This one,' he says, slapping a trunk.

It's hard to believe this is all happening so quickly.

The gate at the crossing point is unlocked and flung open. The ground either side of the old railway is covered by a wet, orangey sawdust skin. A large red storage tank has been lorried in and sits just beyond the concrete block of the old platform. For a good stretch down the track, and two metres to either side, the vegetation has been cut and cleared. Everything has been chopped back, hacked down or dug up. Stumps lie on their sides, brown, wet-rooted with pale egg-yolk hearts sliced open as if to reveal their ages. *Not one of them as old as me.*

There are stacks of branches and brambles, and piles of strimmed nettles. Just beyond the red container is something I've never seen before – it looks like the base of an old lamppost.

The men work surrounded by safety cones and signs. The track has been off-limits for days, but I approached here from the back way, through the wood and over the meadow so I could watch them from the other side of the willowherbs. I feel like I should be here. Someone should be here. Someone should *do* something.

'Don, what about this one?' The young man's voice is loud over the chainsaws. The bush he stands next to is taller than him. Its leaves are spear-shaped and burst out at intervals from its mess of long, arching, flex-like branches. At the tip of each is a large flower cone that looks almost furry. Even now, at the death of August, they are an iridescent lilac-purple; the whole bush looks like a can of paint midway through being exploded. Jets of colour shoot everywhere, too vibrant, too alive, to be hacked down. And that's what's thrown the lad for a minute.

'It's just a buddleia,' yells Don.

'So . . . cut it, yeah?'

'Of course cut it. Then dig its roots. It's a bloody pest.'

Not that it matters now, but it was Bill Varley who first laid eyes on this shrub. Varley who'd been initially unsure as to what it might be (the shoots only had a few leaves on them and his eyes weren't what they'd been), but after checking his *RHS Encyclopaedia of Gardening*, he was convinced. Varley who'd written in his notebook: *27th Aug – I've found Buddleia davidii growing on the dismantled railway!* That was 1990. The species had come a long way since its seeds were sent to London's Kew Gardens from China almost a hundred years earlier. Becoming renowned for its late trusses of colourful flowers and a honey-rich scent irresistible to insects and butterflies, it had earned the colloquial name of 'The Butterfly Bush' and quickly spread throughout England as a much-prized garden shrub. But buddleia wasn't content to be contained for long. Pavement cracks, waste ground, development land, walls, chimneys and shingle banks turned out to be ideal replicas of the rocky screen of its native Sichuan province. By 1922, it had jumped the fence and wild bushes were being reported in all manner of strange and sheer places, rampantly colonising Britain's railway network. Erupting from trackside tunnels and bridges, its rootstock was decried as weakening surfaces and structures and the plant swiftly fell from favour. A black mark was made against its name and rail operators, then government, sought a reclassification. The good-looking, fragrant guest had changed into an uppity, invasive nuisance requiring large amounts of money to control.

Then, during the 1970s, *Buddleia davidii*'s benevolent side once again came to the fore. Not so much pest as provider. A ferocious depletion of natural habitats in and around Britain's towns and countryside was taking a catastrophic toll on wildlife. As meadows, grasslands, farmlands and forests experienced large-scale deterioration, the wild-growing oriental shrub proved to be a vital resource for the rapidly declining populations of pollinators. Conservation charities, initially wary of the risks an invasive species posed to biodiversity, recognised that buddleia's abundance was creating crucial feeding sites for many species of bees, butterflies and moths. And it was the same story in gardens, for despite its demonisation in official quarters, buddleia had never ceased to be cherished closer to home. Being an exotic-looking shrub, a magnet for wildlife and requiring little or no looking after, it had been planted continually by gardeners, including the green-fingered residents of Bilton's new housing estates. Even when fashions shifted in the 1980s and many outdoor areas were paved over with patios, decked, covered by conservatories or turned into parking spaces, buddleia thrived. Residents would find it poking out from what they'd assumed to be soilless spots. And so it happened that when a south-west wind blew through Tennyson Avenue in autumn 1989, it shook a wild buddleia that had burst from a crack in a backyard paving slab. The rustling of its dry flower head loosened a winged seed, which the breeze carried north and deposited in the stony earth beside the old railway.

After discovering the seedling growing there the following year, Varley had kept a vigil as closely as he could, recording its progress among the nettles and wildflowers as they capitalised on a late spell of hot weather. Whether he reported his findings back to the Bilton Conservation Group is unknown. Certainly no surviving newsletter makes mention, and it's entirely plausible that he didn't. The argument over whether buddleia should be eradicated or celebrated had split the room in previous discussions and, in any case, Varley had started to miss the Wednesday-night meetings; rheumatism gnawed at his hips and ankles, and his failing eyesight was a constant source of irritation. He grew tired quickly. Just walking up the road was a painful and potentially hazardous journey. Vanessa worried about him. She felt his rapid changes as if they were her own – the shrinking shoulders, the thinning skin, the breathlessness – and she did as much as she could to offset his frustrations, including painting scenes of the edge-land that she hung in their lounge.

'You just never think your body might become a prison,' he confided one evening as she helped him from his chair. She had managed to hold her small smile until he'd left the room, then buried her face in her apron and burst into tears.

By May 1991, Varley noted that the wild shrub was shooting at an impressive rate and easily outgrowing the competition. Over the following weeks he visited as much as his stiffening, sickening body would allow, inspecting the nettles for signs of rolled-up leaves and caterpillars. It was a gloriously hot summer and when not picking painfully through the edge-land, days were spent resting in their garden, Vanessa painting and Varley listening to the cricket. Their own early-flowering buddleia – *Buddleja agathosma* – crawled with insects and day-moths, ladybirds and peacock butterflies.

'I feel it'll be a good year, this year,' Varley said suddenly one afternoon over the chatter of *Test Match Special*. It was June and England were beating the West Indies at Headingley.

'It will, dear,' she piped back too brightly. 'It will.'

Her chirpiness was an attempt to mask her nerves. Varley's appointment at the specialist in Leeds had been set for ten o'clock the following morning, but as Vanessa told Gladys later at church, she already knew. In the end, she didn't need to hear the consultant's gently delivered words. She could tell from his joyless smile as he guided them to their seats in the carpeted, white-walled consultation room. He wasn't long enough in the job to have perfected a poker face.

Liver, colon and lung. Started in the liver and spread. Quick. Invasive.

'We will of course schedule an immediate appointment with an oncologist to discuss any possible treatments, but I'm afraid it's very late ...'

'Don't worry about that,' Varley had said. 'That won't be necessary. I've had a good innings, thank you.'

It was his 'thank you' that set her off.

He had held her right hand in his all the way home on the train. At Burley Park station, they watched small tortoiseshells and commas swarming the trackside buddleia only two feet from the window. They were drunk with its nectar, sleepy, seemingly unafraid of the great beast that threatened to engulf them. Strange as it may sound, they appeared content.

The edge-land's buddleia flowered that August, flushed with the keenness of youth. Vanessa helped Varley down to the railway to see it. It was the 8th and nature was taking its course. He'd grown too weak to walk any further, but he leaned on his stick and smiled as she nipped back to the car for a couple of folding chairs. Then they sat beside each other a few feet from the shrub – her snapping a roll of photographs for future watercolours, him writing out a few slow, sluggish sentences in his notebook:

It feels like 30 °C. But then my temperature can't be trust [sic]. *All over the place. A woodpigeon heads west. Hoverflies. Cabbage whites. Orange-tips. Jackdaws chatter. Pheasants. I want to catch it all, but I'm too slow. Or it's too fast. Writing is a pain. My hands move at half-speed. My body is disintegrating, although (mercifully) I feel nothing. (Good Painkillers). No pain. I only wish my eyes were still good. So much to see. So much we should all see...*

'Bill. *Bill*,' Vanessa was whispering.

Varley paused, blinked and turned to her. She was sitting, camera lowered. Stock-still.

'Look,' she said, nodding at the foliage.

The nettles pulsed. A blur of green. Thistle flowers. A twist of pale bryony.

'The *buddleia*,' she barely said.

Varley narrowed his eyes and moved them across each of its iridescent flowers. The highest cone moved suddenly as though its tip had taken flight; even his clouded vision registered it. Hovering, it settled on another spike and then opened its wings again. Dark brown wings, almost black in the brightness of the day, save for the stark orange-red bands and white dots at its tips.

Whirr. Click-click. The camera blinked. The photograph would be a wonderful one. Zoomed in tight so that the red admiral and the buddleia filled the frame. Hung on a hospice wall a month later, it was the last thing Bill Varley ever saw.

I've been hiding for a week. Moping is probably a truer verb. The rain has given me a good excuse to stay inside and avoid witnessing the destruction happening down the road. And yet it is constantly on my mind. I've kept pushing the chair back from my desk and beelining for the skylight, flicking its bar and sticking my head out, looking off in the direction of the edge-land. I've strained to hear the chainsaws, but caught only traffic, the beep of a reversing lorry on the ring road, the caged cries of fattened lambs and done-for ewes, and always the digital burble of starlings on the chimney pots and roof ridges. It's evening now and I'm here again, head out of the window, watching the clouds break and bloom like ink in water. Each swirl temporarily exposes the flat, gold of sunset far off west. It seems from another time, like a memory you can't quite recall trying to form itself but forever being thwarted by darker, more pressing thoughts. This whole week my head has felt the same, as though there's something trying to get through to me. But each time it begins to shine through, the anger and frustration clouds over again. All I can think about is how I don't want those men hacking away. How I don't want them interfering, cutting down things that they don't know or care about. How I don't want any part of the edge-land, *my* edge-land, to be ruined. And how I can't do a thing about any of it.

I try to rationalise these thoughts. *Why do I feel it will be 'ruined'?* I suppose I'm just uneasy about the appropriation of places by large organisations, councils or charities, especially when it includes the creation of 'official' pathways through a piece of land. They may encourage people to visit more frequently, but it still feels like a prescribed process. Controlled. Passive. Distanced. Everyone who follows that pathway will experience the land in a very similar way: guided, like a tourist being ushered through. Everyone will see the same trees, the same views, and the same undulations from the same two directions – there and back. It will discourage access to the wider ground and the slower exploration of its space. As a result, it will restrain and deny, creating another form of boundary. The edge-land feels like the antithesis of such management because it has been left to itself. It's always felt abandoned to me, and that's part of its allure. It falls outside the normal governing rules, unaffected by the structured control and design you find everywhere else. Instead, to those who know it, the edge-land has its own tracks, the 'desire-paths' formed by unofficial movements of people like me and the people who came before: Bill Varley and the Bilton Conservation Group, the farmers, the commoners and the millers, the miners and the drovers, the woodsmen and the huntsmen, the lovers and the wanderers. The countless runs of its animals form desire-paths too. All of us have cut our own routes from A to B, crossing the ground in the manner we see fit and not the way a council or landowner wants us to. *It's the control; I hate the control*, I tell myself, shutting the skylight and heading downstairs to bed.

That night a vivid dream comes as I sleep. What my conscious mind has failed to recall all week, my unconscious grabs and bundles into my brain. I see a boy in a bracken field. Fair-haired. Bob-cut. Nine or ten. Wearing shorts, T-shirt and sandals. He is pushing his knees up a rise after his brother who runs ahead on faster legs. It's a hot day and they've already been swimming in the beck higher on the moor. Up where the curlews call out *cour-leeee* in the huge, thick, flustered air. Now, in the late-afternoon sun, the bracken smells earthy, damp, woody, and grows tall as a cornfield. They push through it, cracking stems, sweaty and red-faced with the sun, knowing they're supposed to be home soon, but neither wanting to leave. Approaching their den, they crawl and wriggle through its branched entrance and down into its hole. From here they can see the houses in the valley below. Their house. The edge of a shining town. Then the younger boy shifts on his belly. He looks down at his feet because something is moving past his shoe. He grabs his brother's arm and they both watch in silence as the snake noses up the bank, its little head with piercing black and gold eye, its thick olive-green body and yellow collar. *Grass snake.* I watch them run home screaming for their mother. But I remember it well. Our screams were from excitement, not from fear.

I wake in the morning feeling foolish. And selfish. My memory has revealed some unavoidable truths: the only reason I ever came to know the edge-land in the first place was because that world had been opened up for me. When I came to find this margin I didn't have to learn how to be in it, how to navigate its desire-paths, how to open it up and see its layers, listen to its histories and draw closer to its wildlife; I didn't have to because I already knew how to do all that. Thanks to my mother and father, the fringe had always been a mysterious and wild adventure playground, a refuge, frontier and a portal since childhood. Could I really be against others having the same opportunities? We live in a different world, a changing world. I can't remember how many people I've encountered in the edge-land in all the days I've spent there, but it's not many. And apart from that snowy morning in March when they came to play in the meadow, I don't recollect seeing any children. My concerns over the controlled experience of landscape may be valid, but it's a battle to be had down the line. There's a more pressing crisis. Society is so disconnected from something *real* that perhaps a tarmac cycleway and signposts are required to reach such spaces. At the very least, I suppose, it is a beginning. A way *in*. And maybe as people cut through the meadows or cross the viaduct on their bikes, the pollen, seeds and grasses will lure them deeper; the woods and river will grow in their imaginations. Entranced, they'll wander off the track, cut their own paths and slowly become transformed. It happens to us all.

I look over at Rosie in our bed. She only sleeps on her side now, two pillows under her head and one between her legs to support her hips. She is full-term and suits it, carrying the baby 'neatly' – as the books describe it – at her front, so that you could scarcely tell that she's pregnant if you were standing behind her. Aside from these precise sleeping positions, she is more comfortable than she has been for months. As we get up and fuss around in the kitchen making breakfast, I notice how the bump has recalibrated her spatial awareness. The way she shifts her body back to the exact distance now needed when opening drawers and squeezes past things inch-perfectly, it's as if this shape has always been waiting within her. Inside the safety of its womb wall, our little traveller has come an infinitely greater physiological distance, blooming from a cluster of dividing cells to its final, recognisable form. I try to picture what it must look like now: a tiny assemblage of limbs and fingers and toes, thundering heart, eyes tight shut. Alive but not yet *alive*, cradled safely between Rosie's bones, it is almost ready to emerge.

Through the skylight in my office, an early September day. Beautiful blue skies, long shadows and the air chilled and sharp. Best of all is a vast cloud in the shape of a hare hunkered and hiding in its scrape. I reckon it must be slap-bang above the edge-land. I'll take that as a sign. I open my laptop and search for a word I overheard the tree cutters mention: 'SUSTRANS' – the organisation behind the cycleway. I know the name. It is a sustainable transport charity responsible for the conversion of great lengths of disused railways into foot- and cycle-paths around and about Britain's towns and cities. It is an agency of the edge-land, part of its goal being to promote open access, reduce traffic and increase activity among the unused, unloved areas that fringe our lives. You'd have to be mad to be against that. And as its website resolves on the screen, I'm immediately reassured. There is a photograph of a cyclist, her back to the camera, with sundogs haloing around her. She's cycling away down an old railway. To her right rises the edge of a town; a wall of semi-detacheds sits on a lip of messy, tangled greenery. Beside the track there's bindweed in flower and rosebay willowherb. It looks to be late August or early September, judging by the light. Around now, in fact. The physical resemblance to *my* edge-land is uncanny; it could be Bilton's old railway. Heck, it *will* be Bilton's old railway soon enough. It feels premonition-like – *Here you are, see? Nothing to worry about. Just more people.* And it's true. The ground in the photograph is still wild edge-land; it still possesses the same untamed DNA, it has just undergone a necessary metamorphic shift. An alteration from a solitary to a social existence. A progression essential for all perpetuation and survival.

The homepage photograph fades into another: people walking and cycling to work along an old canal; then another: kids on a bridge surrounded by scrub and high security fencing. It's a close-up and I can see something in their faces: a sense of belonging. This is edge-land as true common ground – reclaimed, re-purposed, nourishing, loved again. And now I feel even more foolish about my resistance to the old railway's new incarnation. The margins *should* be made part of people's daily experience. Of course they should. The wild collision and coexistence of human and nature, the complex interlocking of infrastructure and land, the bizarre and beguiling interchange of the layers of history and modern life – this is what tomorrow's decision-makers must know, experience and understand. This is the reality of our species' interaction with our environment. The edge-land is a visceral reminder that we are all part of a process. It teaches us more about the way things were, are and will be than any grand, aspic-preserved landscape or sterile park. It shows us what we truly are: authors of our own transformations and the transformations happening throughout our world.

The acrid smell of burning drifts across the edge-land again, just like 1979. Not from burning cars this time, but Dense Bitumen Macadam brewed to 200°C off-site and brought, broiling, to the bottom of Bilton Lane. A man flicks a lever on a truck and empties a fresh torrent into a wheelbarrow before steering it along a plank to the rest of his team. *Slush*. Return. They are working fast, taking advantage of this high, warm spell of afternoon sun to sweep this steaming treacle lacquer over the old railway. Another skin, the brown-black track, dries behind them. It is 2.5 metres wide and 60 millimetres deep, rolled into flawless billiard-table flatness, banded on either side by the orange sawdust and turned earth. Just on the edge of this cleared zone, a nettle was caught by a strimmer two weeks previously. Or rather, it was half caught. One side of its thick stem was flayed into shreds, causing the plant to lurch sideways and carry the chrysalis hidden amid its lower leaves from the path of the strimmer's return.

Slush. Sweep. *Slush*. Sweep. The bitumen is spread closer. The smoking tar sets quickly over the ants' nest and the raked soil where the willow tree stood. A trundling roller grumbles past, vibrating the ground. After seventeen days of disassembling and reassembling, something moves inside the chrysalis. The smooth plates of the exterior crack and part. A hunched, black, caped creature pushes itself through the rift. First its left antenna springs up from its stored position beneath its body, then its right. Then front legs flicker out; long legs that give it the grip to drag free its changed body. Resistance falls away. The female red admiral scurries from its now transparent pupa and climbs, carrying its still-wet wings on its flanks, like sodden leather saddlebags.

Slush. Sweep. One of the workmen manoeuvres to get a better angle to brush the macadam and stamps on the nettle, crushing the empty chrysalis. The butterfly holds firm to its leaf until the trembling stops, then continues scrambling up, moving from stem to stem and crossing over into the branches of a blackthorn, warmed and edged by the sun. Underwings folded up, it waits, drying, unrolling its tongue to caress its furry chest, like a dog licking its coat.

Then it opens those wings. Pulsing. Pulsing.

Another red admiral flits over to join it. A male butterfly from further north is migrating south, corkscrewing from ivy flower to ivy flower seeking nectar for strength. For a while they flash around each other and bob upwards, as if blown in great breaths from the ground, the last flakes of a vanishing summer. Then the male is up and away again, climbing higher, tumbling towards the invisible lure of heat on the horizon. The female doesn't follow; it remains basking on the edge-land's blackthorn. And it will stay until March, hibernating through the cold months deep in the pile of cut buddleia stacked beside the cycleway, waiting to be lit by the fiery light of the first warm days of spring. Butterfly, metaphor, indicator species: its presence here testament to the restored health of this patch of earth; its survival through the winter warning of a warming planet.

LAST ORDERS

How long does a nettle live? The thought bursts in. It'll do. He grabs and holds it. Unpicks it. *How long does a nettle live? A year? Ten years? A hundred? More? Think. Think.* Standing at the edge of the wood, half-in, half-out, he breathes hard from running as he stares at a clump of *Urtica dioica*, turning the question over in his mind, forming it into a wall to keep all other thoughts at bay. It has been the same all the way back – Albert, Boulogne. Then Folkestone, London, York, Harrogate. Two days of travelling through the Dear Homeland (a world he can hardly bear to see exists), his attention shifting and fixing on a succession of little distractions. The note of the ship's horn; the rattle of a train window in its casing; the yellow teeth of a woman on the omnibus who, noticing his stare, smiled at him and touched his knee ('we're so proud of you, of all of you'). Her husband's quick, apologetic face and admonishing hiss – *Louise, please.* It is a mental game, a concentration game, an attempt to drill the mind back into proper order by exhausting it in the chase of some innocuous puzzlement. A hope that, if pursued into a fatigued rest, it might sleep and wake restored. Memories gone. Arthur gone. And now the focus is this: a nettle, one of a fresh crop that has sprung forth with the autumn. And all around it the brambles and the pines, 100-foot pines, and clear air, the smell of the river and the creaking of crows. *How long does a nettle live? A year? Ten years? A hundred? More?*

How long will any of us live?

He'd meant to go home. He wanted to try to find that small part of his old self, the small, shrivelled husk of Thomas Watson he'd folded away after 1 July and sent home wrapped in a letter to Elizabeth. He'd convinced himself that if he could just get back to Bilton Hall he might become that man again, at least for long enough to have a bath and dinner. He could only stay a night at any rate. Five days' (enforced) leave meant two to get home, two back, leaving a solitary twenty-four hours in the middle. He'd walked quickly from Starbeck station to the gates of the drive, intending to march right in with a bright call out, like his mother and Evans would have expected him to. Like he used to when returning from Ampleforth for the summer. But there'd been an incident. The little muffled voice outside the butcher's. *Mother! Mother! Look, an officer!* Running feet. Catching up. *Just back, are you, sir?* And blow me, if he wasn't dressed like a Tommy. Khaki, puttees and peaked hat. *We had a Zepp, sir, we did. Over Bilton Lane. Have you seen much?* The boy's face horrified him, morphing as it did into Wilson's, Brayshaw's, Daley's, any of the under-age in his platoon who went into the mud. He'd needed to hurry away. *Concentrate. The game.* The sparrows in the hedge. The oak's outline. The surface of the long road that led him home. The orange sun above the larches. Anything would do. He'd paused to smoke at the end of the drive. The black twisted foliage of the wrought-iron gates had only just been brushed down and there was still the strong scent of soap. Water steamed in puddles beneath. And a handmade sign in black Gothic: *Welcome Home Our Brave Son*, tied with yellow ribbon. The hall was there, just visible. Redoubtable. Unchanged. Dappled by the early-afternoon light and shade. People assembling. His mother would be drifting from room to room, ordering, organising, enacting her role splendidly: Evelyn Watson – the heiress. Insisting hot water was brought up for his bath. Checking the flowers in the drawing room. French doors opened over the lawns to make the most of the views over Spring Wood. *He so loves Spring Wood, you know*. Tablecloths pressed by the maids. The slow *tick-tock* of unbearable time in the drawing room. Cards with Bible readings and sentiments – *Remember, Honour* – propped up against polished silver. The vicar hovering, hands clasped. His sisters. And, of course, later on, Elizabeth and her parents. Sweet, amiable Lizzie who hadn't received a letter from him since August, but had kept up her loyal and loving correspondence all the same. *My dearest darling*... Perhaps Mr and Mrs Hutton would appear at the party's end, shuffling gravely into the room in their mourning dress. Living ghosts, stiff faces trembling, waiting for their moment to ask, their voices trailing away. *We wanted to thank you for your letter. It was a great comfort. Is there anything more you might...? My wife...we'd like to know...*

But what could he possibly say? What could he possibly do? The hall so unchanged and him so terribly. So he'd run from its freshly cleaned gate and hand-made sign, from the Huttons and his mother, skirting the edge of Bilton Park and down through Spring Wood, slowing to a walk by the Nidd. It was the track he and Arthur always took when rough shooting in the Christmas holidays. Walking it again had created a flashbulb of memory – they'd talked of the

woods shortly before going up the line. At brigade head-
quarters, while being briefed over maps, he'd stared at the
four copses intercut by British trenches. The 'Gospel
Copses' as they were christened – Matthew, Mark, Luke
and John. *You see those, Thomas?* Arthur had said, clap-
ping him on the back and pulling on his gloves. *It'll be like
hunting pheasant in Spring Wood.* Arthur's smile, his
fingers working their living way into chestnut leather, his
eager, keen-to-get-in-the-fight face. Arthur's *smile*. The
woods around him. The whistles of birds. And in so remembering he forgot the game for a minute, the device that had
kept such thoughts at bay since the boat over from France
laden with the deformed and dying. And then, by the time
he remembered, it was too late. Arthur had appeared
behind him, clogged in reeking mud. That sucking, sloshing
sound wasn't the Nidd lapping the bank any more, it was
Arthur trying to speak. The trees were thick suddenly with
the stench of him – cordite, shell-soured soil and the iron
tang of fresh blood, like an abattoir's yard. Something had
brushed his elbow. Not a hazel sapling, but Arthur reaching
for his arm. The dead. So many dead. Constituting the
earth itself. That was when he'd screamed and scrambled
up the bank by the weir, praying his friend couldn't follow,
his mind grasping for something else to fix upon. Something
to start the game with again. A brick to build the wall.

He looks at the nettle, noting its small, soft, needle-
covered leaves, the two topmost ones egg-shaped, pointing
up and down, the two beneath left and right. *How long
does a nettle live?* 'Indefinitely,' he says aloud. *How long
will any of us live?* He fumbles for and sparks a woodbine
to try to unclench his jaw, covering the flame and shielding
the glow instinctively (that's how they got Banks – first
night on the line). A deep drag to fog the mind. The tapping

foot. *How long does a nettle live?* Its rhizomes run through this earth, thin yellow roots that, even if dug out, will return. The gardener, Addyman, had told him that. *Pull them out and they'll only come back. Astounding things.* And he'd seen them come back stronger, rising anew from the ground. *Earth to earth, ashes to ashes, dust to dust; in sure and certain hope of the Resurrection into eternal life.* They'd buried as many as they could in the stinking mud behind Touvent Farm. But a fortnight later he'd watched the Germans shell the same ground, blowing the rough crosses and rotting bodies into smithereens. *Earth to earth, ashes to ashes, dust to dust.* And that unholy smell, like a larder full of hanging meat that somebody forgot to gut. *Think, man, think. Concentrate. How long does a nettle live? Where did it come from? It must have come from somewhere? Think.* Perhaps he should uproot it? Perhaps its roots will reveal its secrets? It was his nettle anyway – he could do with it what he pleased. All this land was his, from the hall to the viaduct, the fine gardens, the woods and fields. He knew every scrap and shred of it. Since his father's death, it had come down to him. The eldest son of Bilton Hall. The only son. *You will come home, Tom*, his mother said, proud, but insistent. *You must come home.* The world beyond the fog twinkling. *But what does any of that mean now?* An old life. This had become the demesne of that shrunken, shrivelled him, hidden away in a drawer. His lands were elsewhere. Now he stands at the edge of this place, far away from its inhabitants, as though dead himself. Demented, frightened and adrift. He looks at his hands and sees Arthur's hands feeling their way into chestnut gloves. 'I'm sorry, Lizzie; how could I face you like this?' He is speaking aloud again. 'How could I face any of you? Sorry. I'm sorry. It's best you never know.' And he starts rocking

back and forth and humming. The pines and the crying
crows gather – *aaah, aaaah, aaaaah*. The birds interrupt
each other, yelling out primal, scolding rattles and screams
– *ah...ah...*, *ah-ah-ah* and *aaaaaaah*. So many textures to
their choirs, far and near, like a field filled with the wounded
and the dying. The desperation. The shame of it all. The
shame of his life ended at twenty-one. It was Arthur's
birthday on 22 June. Nine days before the attack. They'd
celebrated together with those two lieutenants from 94
Brigade. *Remember that? The whisky. All drunk at the
estaminet; cigars from the colonel...*

Stop it, for Christ's sake. The nettle. Its square stem dis-
appears into the earth. *Look at it. Clean earth.* He stares.
*Yes. It is good earth. I might have made something of this
once.* A dark brown soil, dotted with insects and pale seeds.
Press your ear to it and you hear ocean. And he does.
Getting on all fours he hears the soft roar of sea, not shells.
He smells it. Cold and pure, like wet wool. And in Bilton's
gentle fields, they are turning it up for potatoes. *Potatoes.*
Under a September sky moving fast with cloud, three
women are wielding forks, turning up sack-full after sack-
full; six more are preparing the soil behind them for a
winter crop. One thinks she spies something at the wood
edge and straightens her back, sweeping her hair into her
headscarf – *I swear I saw someone, looked like a soldier
rising up and replacing his hat* – later she'll say it must have
been a ghost of her Edward, killed at Mons.

They will all be there by now. In the hall, wondering. *Has
the train been delayed? Evans, will you call? They say it ran
on time, ma'am.* He sinks into the undergrowth, out of
sight, and lights another woodbine. The last burned his
palm, unsmoked. He doesn't want to, but he knows he must
look up. Past the clump of nettles the field rises. It is there.

Just at the top. Five hundred yards away. *Serre.* The objective. They are to take it in waves. He can see it plain as day. *It's a fucking fortress*, the soldiers whisper to each other on the firing step. *Stop that.* They are just south of Matthew Copse. An aberration of a wood; nothing left of nature. 'Do yourselves proud,' the sergeant shouts, pacing behind them, slapping packs. *You are Fifteenth Battalion. You are sons of Leeds. Do your city proud. Do yourselves proud...* But there is nothing left of the men, either. Nothing you'd recognise. They are khaki shapes huddled in the khaki mud. Scared boys. Bags of bones and flesh and organs and emotions, loved to death by factory girls and mothers who, miles away, fill their morning kettles as their flesh and blood counts down the minutes. Waiting. Wide eyes. Far-off shouts. Close shouts. Ears ringing now they've blown the Hawthorn mine (heard on Hampstead Heath, he'd find out from Lizzie's letters). A rumour whistles down the line. Rumours. *Dugouts thirty foot underground, steel-fucking-lined.* And he knows they aren't wrong. He saw for himself the night before through the trench periscope beside a worried Captain Haley. The false 'V' shapes in the untouched tangles of wires; barbs as big as nails. *Damn it, Watson*, Haley whispered. The shells had cut nothing. *Lucky if we even reach the wire. Pray God that we reach the wire.* He'd resisted the questions under his tongue: *Why are we doing this, sir? What for? Who for?* 7:25 a.m. Surely his watch had stopped. *How long?* he shouts to Arthur, who is waiting with his section. *Five minutes.* Then, *Good luck, Tom. Good luck.* But still they hold as the smokescreen starts, falters and fails. The crack and rumble echoes along the front. Trench mortars. Grey sky. The dead land. It feels unutterably wrong. Unutterably unnatural. They will never reach the wire. Arthur's thin smile. The whistle at his lips.

For God's sake. Look at the nettle.

The nettle. The nettle. How long does a nettle live? A hundred years? No, more than that. How did he not see before? How could he never see before how perfect and quiet and sublime nature is? He will never uproot another thing – he promises it. Out loud. *The nettle.* He looks at this autumn resurrection of an ancient seed and feels sure it was first carried to this spot in the mudded hoof of a horse. Then the horse appears and he watches it plodding up an old Civil War track that once ran through the wood to the Skipton Road. On the mare's back – *who is that?* Charles I bound and pale, roughly handed over by the Scots and now being escorted from Ripon to Leeds, before his long march to Newmarket. The horse stumbles. Mud and seed are shocked from its shoe. The ground accepts them. The nettle grows and dies and grows and dies for 270 years. Centuries reel past before his eyes. *In sure and certain hope of the Resurrection into eternal life. Could this be real? Is anything real?* He can trust nothing that appears out of the white fog that has been his vision since that morning. The way the soldiers look at him – they must know it too. *You will show character, Lieutenant, or I'll see you on a charge.* The colonel was firm. *Your men need to see some damn fortitude.* What men? *Drink if you have to.* And he'd tried. But then Arthur started appearing from the fog. Arthur as he last saw him. An unbearable vision even drink won't drown.

'You're quite mad,' he says to himself aloud.

The nettle. Think of the nettle. Concentrate. Uro – he dredges Latin learned at Ampleforth – *to burn.* A million little nails on its leaves. The woodbine, again unsmoked, burns his palm and he drops it, and then stares at the circle of shiny, scorched skin it has left. *The nails on the wire.*

They put nails through his hands and feet. First comes crucifixion then resurrection. 'Christ, they crucified us, Arthur,' he says suddenly. All the sons of Leeds that went into that earth. Not a street unscarred from Hunslet to Headingley; not a road of back-to-backs spared. They say the whole city wailed the night the telegrams came. In Bradford too and all across Lancashire. Hooded houses crying through drowsy, blind-drawn eyes. And no son has risen since. Just more and more thickened, sickened, injured skins in every town. New states of existence. New shades of dark. Neither death nor life. The nail through the hand that can't be removed.

He frowns. Somewhere back in the memory – something joyous – there was a nail. Here. *A tree. And Lizzie.* He looks along the edge of the wood. *There!* The ash is only fifty yards away. They'd hammered it there together, a nail low into its trunk, the night before leaving for training huts in Colsterdale, before Alexandria. A kiss, too, down the gully. A fumble above her stockings. Her hair, blond and curled, and her neck that smelled always of lavender. His lip behind her ear. She was everything good in that old world. Such a short time together; little did he know how quickly time would pass. He remembers it all – a day of leave on 24 September 1914. The party at the hall. *The nail.* He wants to see it again but *when did the light change?* He is scared to move. *Their snipers watch for movement.* Dusk has begun to pink Bilton's fields. They must be worried now. Fires lit in the grates. All gathered. Elizabeth's silent tears. Mother's stoicism; the vicar holding her hand. *He will be doing his duty, Evelyn. His duty. He won't want to leave his men.* What men? *Of course, you're right. Evans, will you take the sign down? And tell the maids we'll have the beef cold.* The blackness is on the horizon. He wants to touch the nail, to touch Elizabeth again, but he's afraid. *You don't need to be afraid, Thomas. We'll sign up together. Father says we should. It'll make men of us. There's a jolly tram decked in flags in the centre of Leeds. It's our duty, Thomas.* Arthur's thin smile.

You've forgotten the game again.

Arthur has the whistle in his lips. *Time to go. Good luck, Tom. Good luck.* 7:29 a.m. The grey sky and the dead land. The crack and far-off crack of stray fire. Then deafening, shattering concussions everywhere. Impacts that shock and throw. Captain Haley with his chest blown wide open, blinking and bemused as if merely caught in an unfortunate mix-up with a reservation at his club. The earth bucking in waves, folding in on them the same way brown storm waves smash over the sea wall at Scarborough. The German artillery untouched. Their wire untouched. *Pray God we reach the wire. We will never reach the wire.* Boys swallowed whole by the earth; shelled before they can even leave the trench. How many? Two hundred, three hundred? Count each of them on your fingers. Each one loved. *Out, out. Get up and out, for fuck's—* The sergeant sliced in half. The whistle of the shells. Cordite filling the throat and lungs. The rain of blood and damaged, rancid soil. *Out, out.* The scared khaki shapes scurrying into death, wide-eyed and teeth gritted. Flesh and blood crumpling, as if winded, or bursting apart. And the scream of the shells. The screams of the wounded men. The screams of mothers' kettles boiling on stoves from Hunslet to Headingley. *Out, out. Up and over.* The scream of the whistle still shrilling between his teeth as he runs through the *zip-zip* and hailstorm of burning mud. And ahead, Arthur, on his knees. *Arthur! Arthur! Get up!* But his whole jaw is shot off. He is fumbling, feeling the mess of fractured bone and flesh where his face used to be. Unutterably wrong. Unutterably unnatural. *How silly! How silly you look now in that smart uniform and tie. How silly. How silly all of this is.* Then he is gone completely. Shell and soil dissolve him into a black cloud; they reduce his kneeling form to nothing. *Dust. Ash. Earth. Mist.* And Thomas, hurled back through the air, watches the lines of *zip-zip* over his face like bees in the meadow in summer. The white fog forming over his eyes. 'And do you know what I thought when I was lying there?' he asks, out loud, for he is aware Arthur has appeared again behind him in the treeline, clogged and reeking, rasping and sucking. 'I thought, *Thank God the shell got you.*'

Mmmmmm. Mmmmmm. The noises seem to come from somewhere else. From the ground. From the wood. But Thomas is making them. He is rocking back and forth, humming and shaking uncontrollably. The dead and dying litter the dusk-veiling wood – *aaah, aaaah, aaaaah.* And that sound is not the weir, but Arthur hissing. And Thomas wants to tell him he's sorry and how, after nightfall, he slid all the way back to their lines past those dead and dying and all he thought of was swimming together in the Nidd and shooting grouse on the Glorious Twelfth and hunting pheasants in Spring Wood at Christmas. But he can't speak; he lies on his face, moaning and shaking as the bell of Bilton church begins to toll.

You've forgotten the game again.

Too tired. Too tired.

The nettle. Concentrate. The nettle, for God's sake. The shaking subsides but his hands are still too unsteady to strike a match. *Strike. Strike.* This time he does it and it flares in the air. He draws deeply on the woodbine, turns it inward to his palm. And rocks. And looks at the nettle.

How long does a nettle live? The tobacco rolls in his lungs; the fog, the white mist rolls across his brain as the bell tolls its last. Seven. *How long does a nettle live? Indefinitely.* And he understands that he will never see this place again. This is the end of all of this, of him, of the hall, of Bilton Park, of the world he knew, of everything. That's what Arthur's been trying to tell him. *You don't need to be afraid, Thomas.* But he's not afraid. Not now he understands. Not now he's seen the sunset turning the sky orange and copper and there is the burnt-toffee tang of woodsmoke drifting from the cottages on Bilton Lane. Not now he's felt the *good earth* and drunk in its smell again. Not now he knows the nettle will endure. *There's your eternal life.* And not now he knows he will never be lost in that dead land. 'For I shall leave myself here,' he says out loud. 'All I've lived; all I might have lived. I'll leave it here.' And Arthur hisses in agreement, *Yesssssss, yesssssssss.*

Time to go.

Grasp the nettle, that's what the colonel said the night before the attack. *We must grasp the nettle.* Right now, before that walk along Bilton's fields and under the fire-skies to the railway, before this shell of himself is bundled back through an unreal England and onto a boat for Boulogne. Before he wanders dazed and emptied back up the line to his death, like Charles I trussed on a mare. *First comes crucifixion then resurrection.* He will pass through the veil. The soul into the earth. He will release the edge of his being and let it seep into the stem and slip through its rhizomes, to be locked here in this land. *His* land. *We must grasp the nettle* for the nettle cannot die. *Pull them out and they'll only come back.* How did he never see before how perfect and sublime nature is? He looks at his hand and the pink burn in its centre. 'I will stay for ever.' And Arthur hisses again in agreement, *Yesssssss, yessss.* And so, wide-eyed, half-smiling, Thomas reaches for the nettle and closes his fist around its leaves.

The nettle catches in my palm as I haul my way up the bank. Wrapping my arm around an oak, I feel a sharp sting and shake my hand free. It's a surprising-looking thing, the perpetrator: shoulder-high, gangly, old and woody with twisted stems and frills of yellow leaves trailing to the ground like moth-eaten embroidery. Toppling it against the tree is a weighty cluster of new leaves – as though a hefty sprig of mint has been stuck in its end. They are soft, small and delicate but, as my palm proves, potent enough. The burning takes a few seconds to materialise but I know it's coming. The broken-off tips of countless needles are already embedded, reddening my skin.

I find a ragged dock, crush and spit on a leaf then rub it between my hands, leaving a green, watery stain. Sitting down on the collapsed stones of what was once a wall, I inspect my throbbing hand and then look about me. This is a handsome spot, and one I've walked through before: it is the edge of the wood at the far side of a field where hares run in spring. I'm just inside the treeline, by the seam that separates the mass of pines, ashes and oaks from the crew-cut straws of harvested wheat. This margin is a cascade of nettles lit with the occasional autumnal golds and scarlets of dying willowherb and bramble leaves. Brittle thistles erupt with tufts of down the colour of dirty silk. A bird I can't discern – *a goldfinch?* – squawks a note from high up in the pines. Things crackle, click and stir in the under-growth. It's a blessed place especially now as it gets the last of the streaming sun. With evening falling the light appears to slip down the gradient of the field and puddle at my feet. But there is also a strangeness here, a keen sense of what lies beneath. A thinness in the fabric. This is a margin within a margin. And as the sting becomes a burn in my palm, an odd air of melancholy materialises; I can feel an emotional transference every bit as real as the histamine swelling my skin. There is a very clear sense of someone else being here before, someone else seeing the same views over the field and, behind, down the steep side of the gorge to the Nidd; someone else seeing the lengthening of their shadow and the lowering sun flashing along the same river gorge through the trees. It's moments like these that make you think places have a memory of their own.

It's hardly a theory, more a feeling born of so long spent outside, but what if landscapes somehow become repositories of personal and collective memory? What if traces are imprinted or stored in an imperceptible or intangible way, and the land itself retains the culture of a place? Then, what if when a certain set of stimuli is triggered, a kind of molecular union occurs between that place and a person whereby memories and experiences are passed on like the sting of a nettle? You may laugh and perhaps it's all overactive imagination, but this is what it feels like as I sit and look out tonight from the edge of the wood – the sense of a presence, an emptiness and sadness, not of my making but occupying the ground, as if time is flicking back and forth and beyond worlds, long since committed, buried, forgotten, are leaning into mine.

In the self-indulgent *because you're worth it* tone of our times, the birthing books instruct couples entering the last few weeks of pregnancy to get a few things out of their system. Namely to enjoy all those things that they won't be doing for a while – seeing friends, having dinners out, trips to the cinema or the theatre, late nights and long lie-ins. Your world will soon be shrinking, they advise, horizons narrowing to a small bundle in a Moses basket. The walls of the house will become locked-down borders of your own little, happy, bleary-eyed country where time becomes a surreal concept and a rare commodity. Just finding a second to wash up or go for a shower will feel like an impossible task on top of keeping a tiny baby healthy and happy, alongside the pressures of earning a living. And there are plenty of warnings aimed specifically at expectant fathers: even if you're not the one doing the feeding, don't expect any downtime when you clock off work, mister. You will be in a whirlwind of nappy changing, clothes washing and cleaning up. So enjoy yourself now. Go out before you go in.

Self-indulgence or not, Rosie and I take the books at their word. It's a good excuse, anyway. Autumn is busying itself with politely but firmly evicting the now-destitute summer, apologetically brushing its belongings into the gutters and replacing them with its own ornaments: the dried, browned, straw sculptures of the hedges, road-sides and gardens, the ash keys, the bright, bloody haws and the inky orbs of elderberries; the scatterings of beech mast and conkers. In the wood it has already nailed-up its own curtains. They are threadbare compared to those that hung here a fortnight ago, but exquisite nonetheless, dotted with reds, lemon yellows, greener yellows and all stitched through with the holy light of golden sun. Every night for a week we walk through the edge-land and drink in the long evenings after work, moving at what seems an appropriately sedate pace with plenty of rest stops for perching and

watching. We move at *Pregnant Speed*. Our quiet, slow
tread surprises a roe deer walking through the field by the
holloway; it hesitates and then whips its whole body sideways and runs, stroking the stubble with its hooves as it
passes beneath a pylon and vanishes into the wood.
Swallows on telegraph wires are unfazed as we pass below;
silhouetted against the grey sky, wire and bird become
Franz Kline abstracts of black line and sharp brush dabs
and strokes. Rounding a corner at the top of The Lane, a
thunderclap of riotous applause greets us as hundreds of
little hedge birds burst from their roosts in a cloud flurry,
like flies from a cowpat. They bounce away along the hawthorn and blackthorn, barely touching it, the way a finger
tests a red-hot surface. In the fields we watch rooks rise and
mass in the sky like iron filings around a magnet, pulled
westwards over the viaduct to settle again down into the
rusting trees.

On Friday at five o'clock I'm outside a pub carrying a
round of drinks through a sun-flooded beer garden. The
struts of the bare pergola above me turn the ground into a
grid of light and shade. Rosie has driven south for the
weekend to visit her sisters and I'm catching up with an old
mate, Matt. He has plans for dinner with work colleagues
in town, but there's still plenty of time for a pint or three.
It's always the way with friends you've known for ever: you
see each other far too rarely, and then when you do, you
wish you had longer. We've hardly spoken since I moved
north despite the fact that he only lives half an hour away,
so we drink too fast, trying to make amends for lost nights
and missed meetings. More rounds are fetched. And more.
The light shifts and shadows stretch; we hop tables to keep
in its warmth. Talk turns from babies to what's been occupying my time. I babble about the edge-land and its

extraordinary variousness, about following the fox and the owls and the hares, about Sir Hare and Bilton Spring, about the deer leaping over me, the stoned kids and the mayflies, about the swifts gathering at the sewage works and the butterfly landing on my forehead in the meadow. I tell him about the edge-land's fragmented human histories and try to bring to life its topographic delirium. 'Here, here,' I say eventually, pulling out a notebook and flicking through to the first map I drew. I push it across the table. 'Do you know where I mean?' I run my finger along the old railway. 'They're putting in a cycle-path here.'

He sips his beer and sets down the glass. 'Bilton? Yes, of course I do.'

Of course he does. Stupid question. He is a surveyor in Leeds; he knows the geography of this region as well as anyone.

'You know that whole area's a developer's dream, right?' he says.

'It *was*—'

'No, no. It still is.'

'But they stopped building at the old railway in the nineteen nineties.'

'They did, but they've been itching to re-start residential development ever since. There's a lot of land around there that isn't Green Belt. It's Green Field.'

'What's the difference?'

'One is protected. Well, sort of. For now, at least. The other isn't.'

I spin my map back around and tap my finger on its sketched meadows and fields. 'You mean like these?'

'Exactly. But others too. It's much wider than that. Thirty hectares or so.' He draws a diamond shape with his finger all the way over to the sewage works. 'This area's known as *The Bilton Triangle* and they've been looking at it for decades. It's just the same story as everywhere else in Britain. Massive demand for housing and little available space to build on. But they're trying hard to push things through.'

'How hard?'

'Well, not long ago Harrogate District proposed to allocate land for three hundred houses every year up to twenty thirty.'

'Three hundred a *year?*'

'For the whole district. But the planning inspector rejected it out of hand.'

'I'm not surprised. That seems an incredible amount ...'

Matt frowns at me over his glass. 'No, mate. They rejected it because it wasn't enough. The developers are suggesting nine hundred a year.'

I feel stricken. I try to imagine what that would look like, but it scrambles my brain. 'Hang on. So what's stopping them building over it now?' I ask. 'Local objection?'

He shakes his head. 'The roads. There isn't the traffic infrastructure to support the numbers of homes they're talking about. It would mean a load of work and disruption to build the new relief routes required.'

'But they will eventually?'

He nods. 'They'll have to.'

'How long?'

'Could be ten, twenty, thirty years. Might only be five. But they desperately need the houses so, in the end, it's only a matter of time.'

After checking his watch he pushes his chair up and gathers our empty glasses. 'Speaking of which. One last round? Then I've absolutely got to go.'

But he doesn't. And neither do I.

As night falls we repair to the main room and grab seats. After all, I'm under instructions. *I'm supposed to be here*, I remind myself while stocking up on more of the locally brewed ruby red. Then before we know it, a bell is ringing and the barman is shouting to make himself heard. 'Last orders,' he bellows. *'Last orders.'*

All the way home along Skipton Road, walking unsteadily at *Pregnant Speed*, I think of the irony of the edge-land being safeguarded by a road, and of Matt's words – *It's only a matter of time.* I suppose you could say the same for anything. For the universe. When you boil everything down, all is a matter of time – or matter *and* time – but some things are too large to comprehend. I never expected the edge-land around Bilton to remain as it is for ever, I just hadn't thought of it ceasing to exist at all. We're not good at confronting endings until we're forced to, even though the unshakeable truth of our own conclusion is always with us, that deep terror lurking in the back of the mind. That curse of our consciousness. We become masters at training ourselves to forget about it, concentrating instead on the many joys and wonders of living, and the little distractions and the destinies we feel we can control – jobs, relationships, family, friends, going to the pub, the routes of cycle-paths. These are the walls we build to keep those other thoughts at bay.

Crossing a bridge over the train track my eyes are led off by the row of arc lamps below, a diamond necklace of corollas strung through the darkness towards the distant edge-land. Around me the houses thicken in every direction, traffic whooshes past, streams of taxis and buses pour out of the town, headlights blazing, sweeping me with their tunnels of light. This spot was once an edge of town itself. It too was woods and old paths and memories before it was commandeered by the sprawl for an intersection. There's old man's beard growing along the railway, elder and ash reaching over the bridge's sides and the pavement. A pine curls up from an isolated square of grass outside a school. Through blurred eyes I can just about picture what it must have been like before. But in this place I also see what the edge-land will become. Time flicks back and forth like the cars. Beyond-worlds leaning in again.

Sometime around mid-afternoon the hangover slackens in its ferocity and I roll myself off the sofa, switch off the TV and head upstairs to pack a rucksack. With Rosie not returning until tomorrow and that creeping sense of alcohol-induced claustrophobia, I know where I want to spend the rest of the day and the night. Part of it is about clearing the fog in my mind, part of it is making the most of the weather – the forecasts are adamant: *Enjoy it while you can for we're in for some REAL autumn weather next week with high wind and rain sweeping across the country* – but these are only the surface reasons. The truth is that I want to sleep out in the edge-land again – now, while I still can – and remember it all.

Swinging my pack onto my shoulders, I take the long way round because the sky is as beautiful as I've ever seen it. From our street, above the chimney pots, aerials and pigeons, silver-spun clouds roll like surf breaks. Over the ring road starlings swap loud swannee-whistle notes from rooftops and the crowns of pavement-locked lime trees. Behind them the expansive plane of shape and light continues unheeded. In one direction it is the soft-focus of a Renaissance painting, tinted as if to emphasise some important event happening on the ground below. Southerly it is more Dalí – flat, crisp layers of blue and the steam train billows of cumulus. The sun is hoisted high and hot on one side of me; on the other, above the rotting fascia of a William Hill, is a trimmed toenail of moon.

At the crossing point, the cycle-path looks smart and flat and tidy. It is the same thick black as the bitumen cliff drooping down the raised concrete block of the old coal platform for the Barber Line. I can hear the diggers clanging away out of eyesight. They've almost finished tarmac-coating the viaduct already, pushing on over the Nidd with the same speed and spirit as the Leeds and Thirsk Railway Company before them. 'The Sailor's Hornpipe' blares dementedly from an ice-cream van somewhere in the Poets Estate. Thanks to the cut-back trees and shrubs I get a view of the fields beyond and the fuzzy green backbone of the holloway. Clouds form the shape of a basking shark, mouth open, drifting over the pale land.

I've had this annoying phrase flicking around my head all day: 'Bilton will be *built-on*.' The clue was in the name all along. It seems so obvious now, both the wordplay and the inevitability of the matter. I mean, what urban fringe won't be built on in years to come? And how can we object to it anyway, we who have babies on the way? I'm bringing new life into this overcrowded world. Where can I reason-ably expect our children and our children's children to live? Everyone involved in the debate – politicians, planners, even those who object to new developments – all agree on one thing: the housing crisis is only going to get worse. The evidence from demographic trends suggests that six million new homes are required over the next thirty years. That's 200,000 per year, and yet, in England, building just over half that number in the last twelve months has been fraught with problems and setbacks. I know all this, of course, and

I've picked at these thought-threads a thousand times, but now I've found this place, I hate to think of what will be lost, of the monumental shift that is coming even for these long-humanised lands. I know Matt's right; it is just a matter of time. Nothing lasts for ever. Not me; not this land. Bilton will be built-on, but it hasn't been yet. The road still holds the future at bay. There's still tonight, the wood, and breath in my lungs and maybe, in the end, that's as much as we can really hope for.

'There was nowhere to go but everywhere,' wrote Jack Kerouac, and that's how it feels as I meander down the holloway, pausing to sit and marvel at the mystical light slipping in strips through its woven hazel walls, barring the stones of the drovers' track. I don't need to slow my pace now, to rest frequently, but I think I've grown addicted to moving at *Pregnant Speed*. The hawthorn hedges are washed with a silvery shimmer; their leaves glisten like fish scales. Flashes of flame leap over me between the crab apples – flitting chaffinches and robins – and I spin my head around, trying to follow their light. The sun stays aloft but becomes oddly heatless. As I pass into the wood, the air feels colder still, so I pull on my jacket then roll along upstream beside the river. *Knock-knock-knock*, a nuthatch taps a haul rhythmically against an alder branch. Sweet cicely grows in parsley-like clumps and I grab great handfuls and breathe in its school-bus scent of aniseed. Away to my left, on the ridge of the high bank, there is a crash loud enough to be heard over the static rush of the weir. The profile of a roe deer bounds through a holly bush. I watch it disappear. The rush inside me fades and I make my way up the gorge side, using the tree roots and saplings as handholds, keeping an eye out for the nettle that stung my palm last time.

At the top of the bank I slide off my pack and fix up my hammock, threading its heavy cord around the trunks of two sturdy pines. This is where I've wanted to be all day. It is this handsome, haunted spot that lured me from the sofa. This small space just inside the treeline where I can watch the last of the light slip down the field and linger in the nettles. This thin place in the fabric, this margin within a margin. Where day meets dark. And after a few minutes I feel the same sense of correspondence as I did before – the same sense of presence and the awakening of memories. I feel a resonance with the thoughts and lives and layers committed to this ground, as though faintly remembering an old dream. A calm, chilling intimacy. A knowing. And I'm not afraid. I'm not afraid of anything. On the contrary, in fact: I came here tonight searching for just such company, for this union, for the fellowship that would share in and make sense of my own feelings of needing to say some kind of goodbye. The going out before the going in. Perhaps to slip through the fabric and leave a trace of myself, immune to age and forgetfulness, consigned to this precious, doomed land.

Force of habit: I wait another hour before lighting a small fire. I haven't seen another person, but I don't want to have to leave now. I couldn't face that. Not as the excitement of the slow autumn dusk descends and the bliss-nerves that come with sleeping out alone swell in my chest and crawl up the skin of my arms. Dusk is the moment when the simple act of spending a night outside seems thrillingly transgressive, somehow an inverting of the orders and strictures of our everyday lives. The senses are sharpened. Ears strain. Eyes watch for the slightest movement. You are suddenly very alive. Searching for firewood I pick my way along the margin, through its million stings, gathering up the standing dead, snapping branches that have slumped but not yet rotted and piling them around my camp in bundles. Woodpigeons clatter up from the field. A crow barks. Squirrels scramble around the trunks. Water in the ruts of a tractor's tyre trails are deep lagoons, slicked an oily bronze by the sky. Amid the brambles and nettles, rosebay is feathering at its ends; everything exudes the pungent, ferny tang of leaf mould and bracken. Near the hammock, on the field side of the collapsed wall, I make my fire, sparking the shreds of honeysuckle bark with a match and watching them twist and turn to become whining, flaming tongues that lick the pine kindling into life. Over the field and the wood and the pylons, the post-sunset sky gleams above the hills with the same fiery tones: bursts of yellow, bright white and the ash-grey clouds. Geese honk westwards in a drifting curve, lifting my attention up to the migrant skyways, the invisible songlines that now ripple with the voices of departing birds. After a while others intersect them: the calls of birds returning for winter. Wind farms spin on the furthest skyline behind the town, glowing red lights at their propeller centres. In the sprawling space between here and there, streets echo with dog barks and sirens; columns of smoke rise from chimneys like a forest of ghostly trees. A bell rings faintly.

I stack the fire for warmth and cook my tea – Super Noodles – on a little gas stove close by. I'd meant to buy a tin of vegetables, but forgot. No matter. I put on my gloves and pick a few young nettle heads, adding them to the bubbling broth, watching as they turn the dark green of spinach. I eat it all mixed together straight from the pan as the blue-black twilight snuffs out the last of the day, dissolving everything. Dark now. I hear noises register over the sound of the weir: footfalls and brushed foliage. Things shuffling, gathering, dispersing. From down in the gorge, the trembling shrill of a tawny owl. I load up the fire, then unpack my sleeping bag. The hammock has a flysheet in case of rain, but I unclip it so I can look straight up through the pines at the scattered stars. And here I lie, warm and cocooned, with only my face exposed, suspended between the endlessness of the universe above and the thousands of years beneath. Absorbing everything, I drift off with the sensation that I'm descending into the blackness of the gorge, being absorbed deeper into the earth.

REVELATIONS

Seeking a voice in the dark, I flick on the radio and walk straight into an argument. An impassioned affair, already well underway, echoes off the walls, forcing me to dash back and turn it down. *That is not what our research suggests* ... growls an irate government minister before the interviewer cuts in: *But it is what the evidence says and isn't that the problem? The science is clear. This cull will not have any significant impact. It may even make things worse* ...

I rub my eyes and open the cupboard. The oven clock glows 6:51 a.m. Between that and the radio's display there is enough light that I don't need to switch on the main bulb. Black presses against the kitchen windows, invading the corners in the cold sweat of condensation. Arms folded, leaning against the sink, I listen to the debate as the kettle bubbles up to boiling, but it feels too early for such barking, especially when it's going nowhere. The minister refuses to concede ground and the more it escalates, the more he begins to sounds like a school debating champion entrenched for the sake of it, bloody-minded and priggishly fighting a line he's been handed. I start to wonder whether even he believes what he's saying. As he begins yet another sentence with *Let me be absolutely clear*, I hit the 'off' button. Enough with the professional politicking. It's been the same for weeks now and the only thing the government is not being is clear. *Clear* as in unobstructed; clear as in transparent or without impurity; clear as in evident to the mind, free from guilt.

Sipping coffee, I stare through the window into the shapeless gloom outside, scanning the pre-dawn dark behind our house for sunrise in the east. It's taking its sweet time but then again, that's October for you. All hush, no rush. The slow-start days. Murky, misted mornings. The steady unpicking of the trees. The clocks going back. And in keeping with the rhythm of the month, our baby is now nine days overdue. It's a strange feeling that comes with the waiting, like we've momentarily fallen off life's merry-go-round into our own excited space. We're on permanent standby and entirely freed-up. Appointments are cancelled. Meetings re-scheduled. I work when Rosie sleeps – in these early hours, an hour or two in the afternoon and later on at night. The rest of the time is given over to enjoying our suddenly spontaneous days. It's like going back to when we first met at university, the impulsive, hour-by-hour existence of students newly let off the leash. *Let's tramp the moors; let's go get ice creams; let's get the train to the coast!* There is a freedom and intensity that comes with knowing that things won't stay as they are for long. A strange simplicity; a clarity of sorts.

It was a snap decision yesterday to head out for a walk through the edge-land, coming in from the west and cutting up through the meadows to the old railway, following the cables threaded from pylon to pylon. It was getting on for late afternoon but the sky was still warm and huge, the sun a luminous disc ringed by a circle of pale blue. We ate the last of the blackberries and sat for a while sky watching, eyes shifting from rubbly clouds with gilded edges to great chrome-coloured sweeps. Rosie suggested we walked a loop, down the holloway to the river, along the gorge and back past the sewage works. Messing around where the track sinks into the trees from the field level, I scaled the slope and kept pace with her as she descended along the woody tunnel. The soil beneath my boots was like cocoa powder, recently milled to a fine tilth. It stretched out to the straggly hedges to the west and south and, where the earth curved, it took a red tint, like dried blood. The prints were obvious in the surface; a series of darker, round dots, but the soil was too soft to retain their details. I called Rosie to join me and, as she cut through the trees, she found the others – a whole channel of them pressed into damper ground at the edge of the wood. Hind and fore paw prints overlaid each other – rough ovals, like soap dishes, with traces of five-clawed toes. 'A badger?' she asked and I nodded. More than one, in fact. Prints of different sizes. Possibly a family group. I snapped a photo of the clearest, then fit my hand in its depression, scrunching it up so that my palm sat in the pad and I could push a couple of my fingertips into the holes left by its toes. It suddenly mattered to me, the act of touching those traces. It mattered because I'd seen neither hide nor hair of a badger since January and now, here, was the irrefutable proof they were still around. It mattered because the way they wove between the borders of the 'wild' wood and the neatly ploughed field reminded me of the edge-land itself – unseen, evasive and surprising. And it mattered too because those muddy prints traipsed an unexpected darkness into our expectant world; they were a breath of wind down the neck. They made real the madness of what was happening in the outside world, giving a shape and a form to a shy creature currently being dragged into the spotlight and fought over in every corner of the national press.

The badger (*Meles meles*) seems, on the surface, an unlikely candidate to wind up at the heart of any media frenzy. Nocturnal, redoubtable, clean-living, enigmatic, rarely seen at all other than as a dead black and white bundle by the side of the road, it has only been thrust centre stage at all because of our reliance on, and increasing exploitation of, another animal – the cow. Britain's ever-squeezed and long-suffering cattle-farming industry is stuck in a dirty war with an intractable disease. Bovine tuberculosis is a ruthless and, according to recent government data, resurgent virus that causes tens of thousands of beef and dairy cattle to be slaughtered every year in England and Wales, striking with particular virulence in the south-west counties of Gloucestershire and Somerset. 'The White Death', as it's known, devastates wherever it hits, exacting a heavy emotional and financial price from farmers as it jumps through their herds and across neighbouring lands, wiping out prized bloodlines, forcing the sudden shut-down of livelihoods, restricting movement and heralding the precautionary killing of huge numbers of beasts, many of which turn out to be disease-free in post-mortem. It's a crushing process for families to have to bear and a terrible psychological burden to live with that constant fear. Testing has become more rigorous and regular but even half an eye on the news will tell you that bovine TB continues to be a thorn in the side of the 'rural economy', stacking up a costly bill for the British taxpayer as a result. Those thousands upon thousands of prematurely killed cattle have incurred a half a billion pound bill paid to farmers in the ten years between 2001 and 2011, mostly in compensation. With the government's Department for Environment, Food and Rural Affairs (Defra) estimating that amount will double over the next decade, the pressure to win the war on TB has become intense from all sides.

The problem for the badger is that, just as the virus jumps from cow to cow, it is also passed on to other species through breathing and exposure to cattle faeces and urine. Sheep, pigs, cats and deer can all be infected, all could theoretically bring the disease back into a herd, but the badger has long been suspected by farmers as the most problematic of these carriers because of its home and habits. It lives in the woods at the edges of fields, often sharing the same ground and foraging at night among drowsing stock, industriously toeing over cowpats and digging earth for worms. In 1971, when a dead badger tested positive for the bacteria in an area of Gloucestershire badly hit by bovine TB, the Ministry of Agriculture made no secret of its results and almost overnight the animal became a national scapegoat, despite there being no hard scientific evidence that badgers were even capable of transmitting the disease back to cows. A villain had been caught red-handed and long-held suspicions about the 'wildlife reservoir' being at the root of the re-infection of cattle appeared vindicated. A causality link existed, no matter how tenuous, complex and multi-faceted. It was enough to herald a new front in the war on TB: the first official culls of badgers in Britain; the mass gassing of setts near hot-spots. Hydrogen cyanide was supposed to kill gently by putting the animals to sleep, but when the gas was shown to cause lingering, excruciating deaths, it was replaced with the swifter and 'cleaner' process of trapping the badgers live and shooting them in the back of the head. Throughout the

1980s and '90s the link between badgers and bovine TB levels in cows was tested and retested through mostly unscientific field studies and culls. Trapping and shooting were licensed in a reactionary move wherever outbreaks struck, provoking little fanfare or public controversy because the numbers of cattle and badgers killed in each incident was relatively small. Figures collated since the 1980s showed that both bovine TB and badger numbers appeared to be on the rise, but actual evidence of the animal's contribution to the disease remained, at best, inconclusive. After forty years of policy failure in tackling the virus, there was a large-scale independent review of bovine TB management policy in 1996, chaired by Professor (and later Lord) John Krebs. This recommended that the government undertake a scientific trial to establish once and for all whether badgers were responsible for a resurgence of bovine TB in cattle and, importantly, whether killing them would help reduce the incidence of the disease in stock.

The Randomised Badger Culling Trial (RBCT) was conducted over nine years between 1998 and 2007. It involved the deaths of 10,979 badgers at a staggering cost of nearly fifty million pounds but, despite opposition from animal rights activists, it was widely held as money and time well invested. This was the longest and most in-depth experimental study of the effects of badger culling on TB in cattle in the world. It was what Defra itself called 'the best scientific evidence available from which to predict the effects of a future culling policy'. The RBCT found that, at the most, culling would reduce incidences of cattle TB by 12–16 per cent over nine years and subsequently, in the words of John Bourne, the scientist who led the trial: *badger culling cannot meaningfully contribute to the control of cattle TB in Britain. Indeed, some policies under consideration are likely to make matters worse rather than better.* The Labour government followed his recommendations, but it was clear even before publication that not everyone wanted to listen. During 2004 the (then) shadow environment minister, Owen Paterson, was busy tabling a record number of Parliamentary questions (almost 600) on bovine TB. He was keen to bolster his image both as a 'true blue countryman' dedicated to eradicating the disease and – prudently, in political terms – as a willing mouthpiece for the powerful farming lobby, the National Farmers Union, making it clear he shared its belief that badger culling was a necessity, regardless of what the trial found. So when the coalition government took power in 2010 it came as no surprise when it made good on its promises, revisiting the RBCT trial data and, inexplicably, drawing a different conclusion to the scientists: that killing badgers in substantial numbers is essential to stemming the rise of tuberculosis in British cattle. Suddenly culling was back on the cards.

In recent weeks, and with Paterson freshly appointed as environment minister, it has been rampantly pushed up Defra's priority list. The pilot shooting schemes are due to begin any day now in west Gloucestershire and west Somerset. But this is only the start of it. If 'successful', the government is committed to rolling out the cull nationally, which could mean many tens of thousands of badgers being killed, regardless of whether they are diseased or healthy. This despite the fact that the cull goes against all rigorous scientific evidence and economic logic and, if you believe the experts, will have no meaningful impact on reducing bovine TB in cattle.

The government must surely have expected a backlash – this is, after all, highly contentious policy – but even so I doubt many in power could have predicted that the plight of the badger would become the inflammatory issue it has among British voters. The opposition is growing daily. No wonder the minister sounded so indignant on the radio this morning. A public e-petition calling for the abandonment of the cull and a Commons debate on the issue has amassed over 150,000 signatures; rock stars, famous naturalists and spokesmen from wildlife charities are appearing on TV shows imploring the Prime Minister to intervene; police are warning Defra over spiralling costs as activists pledge to disrupt gunmen and frighten off targets with a chorus of vuvuzelas left over from the World Cup. Images of badgers are appearing on every front page; they are colonising the streets and newsstands. But amid this tragicomic cycle of events remains the sobering weight of informed opinion stacking up against it. Opposition voices are pointing out that the cull is a distraction from the real war against TB and that the money could and should be spent on the host of other more effective strategies – biosecurity and further developing and employing TB vaccinations for badgers and cows. In a body blow to the new environment minister, thirty eminent British scientists who work in animal disease treatment and control have sent an open letter to the *Observer* insisting that the government rethinks its strategy immediately, writing that the evidence shows the planned cull may actually risk increasing TB in British cattle. This is because the proposed free-shooting approach (i.e. culling without trapping the badgers first) could encourage the 'perturbation effect' with badgers roaming from setts into wider areas, potentially spreading the disease into other herds. Under this intense scrutiny, the cull's flaws are revealed. Free-shooting badgers is a cheap option compared with trapping and killing, but it is distinctly unscientific and without control areas, testing or the possibility of

statistical analysis. As such, there can be no way of measuring its effectiveness or even determining what percentage
of badgers killed has bovine TB. Campaigners claim this is
because the data would corroborate the findings of the
RBCT and show a low incidence of the disease in badger
populations. Most troublesome for Defra, however, seems
to be the challenge of meeting the cull's own objectives –
namely to kill the requisite minimum of 70 per cent of
badgers in the pilot areas. That would require knowledge
of how many animals existed before you began and it turns
out that nobody has that exact information. Lord Krebs,
architect of the RBCT, describes the whole plan as 'mindless', accusing the government of cherry-picking data to
justify an agenda. Lord May, a former chief scientist, agrees,
saying: 'They are transmuting evidence-based policy into
policy-based evidence.' The public is rightly suspicious of
this all too frequently employed political inversion; it
smacks of the tactics the previous Labour government used
in the trumped-up case for war in Iraq. And yet the more
farcical it all sounds, the more ministers are becoming
badger-like, backed into a corner, digging their heels in,
bearing their teeth. *There must be a reckoning*, they insist.
And soon.

'Maybe you are being too sensitive,' a friend said to me
the other day as we talked it over. 'You've grown soft. It's
understandable; you're having a baby any minute.' He was
being flippant, teasing, but his words stung me. 'Sad senti-
mentality' is an argument that the cull lobby has been quick
to smear the opposition with, citing the traditional borders
between the removed, sanitised mentality of the town and
the harsh but necessary realities of those who deal in
country life. But I've known both these worlds, and I've
killed animals before, shooting, butchering and eating

rabbits and woodpigeon that threatened to overrun a farmer's crop fields. I've felt hearts slow and stop under my fingers and seen eyes set and staring because I chose to pull a trigger. I've watched emaciated red deer stags, ribs poking horribly through fur on a Scottish mountain, starving to death from overpopulation and the human-wrought absence of predators, and I'm certain I would have shot them there and then too, had I been carrying a rifle. I've broken the necks of rabbits squatting swollen and hopeless on footpaths to end the unholy misery of myxomatosis. No, I don't think I'm squeamish; it's true that I've never managed to stop my heart hammering in my chest or shake the repulsion at my own hands for hastening the end of another creature, but that innate displeasure has proved an ethical compass too: I've never killed anything for the sake of it. There's always been a reason. That's what I can't shake about all this mess, that there is such a lack of logic in the sanctioned mass killing of one of Britain's most intriguing wild animals. It's proving damaging for the coalition too. Dragging the badger into the line of fire has had the unintentional outcome of hauling the workings of the government and politicians into the crosshairs. I know I'm not alone in wondering, *If the government is refusing to listen to science and reason, who or what is it listening to? And why?*

'Let me be absolutely clear ...' the minister had said on the radio. Well, you should be careful what you wish for.

It's still dark through the kitchen window as I pour another coffee and cradle the cup's warmth in both hands. The coughs of starting, stirring traffic sound faintly from the front of the house. I think of Rosie sleeping upstairs and a baby, like the dawn, on the edge of arriving in this world. *But what world are we bringing you into?* It's a thought that has bitten into me many times recently and I feel it sharply again now. There is a consciousness gleaned from time immersed in the edge-land, from being outside the confines; a perspective that comes from being where the historical collisions between human and nature are evident and inescapable, like a movie permanently projected along the borders. Playing on a loop, it whirrs back and forth, showing us what we are and how we came to be thus. The kitchen radio is switched off, but the minister's barking rhetoric still echoes in my ears, jarring and unsettling. His talk of growing 'rural economies' and making them ever-more profitable smacks of a wider addiction to endless growth that doesn't square with the finite resources of this biosphere. There is a debt accumulating that no one wants to address. In its current state our economy already consumes 50 per cent more resources and churns out more waste than ecosystems can restore or absorb. So what will happen if and when that does increase? I think of the phrase I heard a different minister say on the radio many months ago – *anything that cannot justify itself financially has to go*. The second, unspoken part of that being: *anything that can justify itself financially is fair game*. Now, after these many months, I can hear in those words the same cries for 'improvement' that justified the acts of enclosure in the eighteenth century and, later, monoculture and the intensification of farming. I sense the residual influence of a wealthy minority that has governed this land in various guises since the Normans and whose motivation still appears to be self-interest, self-preservation and power. I feel sorry for farmers at the sharp end with ever-increasing overheads and pressure to produce more, and who are already choked by a consumer society that demands cheaper food yet throws away 30 per cent of what it buys. I feel sorry that they're being driven out of business, ignored, lied to and manipulated, that we all are. And I feel sorry for the other species caught up in it all: the cows bred by artificial insemination for fifty years to yield more milk but that are now weakened against disease; the golden eagles killed to protect industrially farmed pheasant for sport; the pollinators being poisoned by pesticides; the badgers about to be shot for nothing more than political purpose. These things may only be local and small-scale but they form part of a bigger global picture where extinctions, habitat loss and climate destruction are raging unchecked, and where almost half the planet's population of invertebrates has been lost in the last four decades. Maybe I *have* grown sensitive, but only if sensitivity is waking up to the state of things and having a more acute understanding of, and empathy for, the intrinsically valuable and miraculous world we exist in. A world that science and rational thought tell us we're pushing to the edge and well past the point of no return.

And standing here alone looking into the blackness, I'm struggling with even blacker thoughts. There is that niggling, inconvenient question refusing to budge: *If the government is refusing to listen to science and reason, who or what is it listening to?* The news of the appointment of Owen Paterson as Secretary of State for the Environment is not merely a death sentence for thousands of badgers, but a cruel and cynical twist from a government that once pledged to be 'the greenest ever'. Paterson is an avowed supporter of fracking to extract shale gas and an opponent of renewable energy subsidies but, most incomprehensibly of all, he is an open climate-change denier. In the name of 'global trade' he is making promises to tear down any environmental red tape around big business, burn more fossil fuels and support the expansion of the UK's airport capacity. Yet the government *knows* the physics is unanswerable: climate scientists agree that carbon emissions are dangerously heating up the planet. The Intergovernmental Panel on Climate Change warns that most of the global temperature rise since 1950 is due to greenhouse gases and deforestation, and that a further increase of 2°C would be catastrophic for humanity. There is no debate; eminent scientist after eminent scientist agrees that it is our consumption causing the unprecedented ice melt and global temperatures and sea levels to rise. The proof is everywhere that we are beckoning catastrophe and yet he, the environment minister, is refusing to meet with the chief scientific adviser at the Department of Energy and Climate Change. It is like watching a car crash in slow motion. It just makes me think, *Why? Is there something inherently wrong with us? Is there something immoral at the core of our species?* These are dark days when those elected to run our world pour scorn on scientific consensus for short-term gain and to protect the interests of mining, oil and gas corporations. Dark days when the ultra-rich beneficiaries of endlessly increasing global growth know the consequences of continuing to produce carbon emissions, yet invest hundreds of millions of dollars to clandestinely influence political agendas, lobby government and finance 'think-tanks' to rubbish climate-change data. Dark days when you realise you are bringing a life into a world that you're not even sure you can trust any more.

When the first liquid grey light steals through the yard, it takes me by surprise. I blink and peer out at the familiar shapes forming in the silver-black – the log shed, the little flowerbed, the back gate, the trellis coiled with a climbing rose gone wild. Eastwards a hairline crack in the black has opened above the rooftops. It is fascinating thing, vanishing to a tiny point like a bronze road into another dimension. Then, from above, a muffled cry breaks the silence. It comes again, clearer now. Rosie is calling my name in a voice I've never heard before. I drop the mug in the sink and run, taking the stairs two at a time, but before I even reach her, I know what it means.

For the past nine days we have had a hold-all packed and ready, stashed in the corner of the bedroom by a Moses basket, similarly primed: cleaned, blanketed and waiting for occupancy. In our 'hospital bag' is an array of oddities to ease the journey ahead – flannels, drinking straws, an iPod filled with relaxation music and affirmations. And, sitting on top, my tatty notebook. After a few hours the contractions have quickened in frequency but our house isn't very far from the hospital and two phone calls later, we're still at home. Still waiting. By ten o'clock all that remains for me to do is to make sandwiches. *Mum will need her strength* we're told in the NHS leaflet, also tucked into the bag, like an invite to a party. Even so, it seems an absurdly mundane task when every six minutes, counted out carefully on the oven clock, Rosie is doubling over, gripping my arm and riding a sea-swell of internal pressure for sixty seconds, sighing, breathing, humming. I hold her, support her body and rub her back as she crests each surge, then dash back to peeling the boiled eggs and mashing them up with mayonnaise. 'What about the smell?' she asks, leaning on the sofa recovering her breath. 'Won't egg mayonnaise stink out the place?' I point out there'll probably be worse smells to contend with in a maternity ward and then admonish her for always worrying about others, even now, while secretly thinking how wonderful that is. But we are out of bread anyway. We both laugh. Amateurs; so excited and so frightened. After another contraction passes I dash to Sainsbury's around the corner. The world outside the door is superficially the same – cars and traffic lights; a clear October day, born cold and growing colder – but it feels like I'm on a different frequency, a different rhythm, and numb to everything else. It's like being caught in a tractor beam radiating from somewhere beyond, a warm, nervous energy that is pulling me steadily and unstoppably towards a place where nothing will ever be the same again. It's only when I run to the checkout clutching a loaf that I discover I've left my wallet at home. 'My wife's about to have a baby,' I mumble, 'I'm sorry.' I'm halfway out of the door when the girl on the till calls me back. Her eyes search my face for something. Honesty, perhaps. She curls stray bright-red hairs behind her ear as she checks the aisle behind me. Her purple shirt moulds around a distinct bump of her own. Five or six months, I'd guess. 'Here,' she says, handing me the bread. 'Take it. Bring the money another time.'

It's a kind act, one that could get her into bother, and I'm touched. I read the name badge pinned to her fleece and smile. 'Thank you, Lauren.'

An hour later our bags are by the door, coats on top and I'm pacing the hall itching to do something. Rosie is back on the phone to the maternity ward. The contractions are getting stronger and the midwife is asking when she last *definitely* felt the baby. When Rosie explains it was sometime during the night I hear the voice at the other end click into a different gear: 'OK, right. Well, you need to come in now. We'll have to check the baby's heartbeat.'

There are things I remember from coming here for Rosie's scans: the surprising hugeness of the hospital, its coral-toned brickwork and blue railings, the buddleia now gone to seed in beds by the steps, the warmth of the corridors and the smell of disinfectant. What I missed on previous visits were the two photographic portraits of the Queen and the Duke of Edinburgh staring down benevolently from the walls. Opposite them, so it appears like the Duke is struggling to make sense of it, a map explodes the hospital's many floors and wings into details. It's like a disassembled diagram from a car manual, too much to take in. I just see random words: Radiology, Cardiology, Chapel, Restaurant, X-ray. 'It's down here,' says Rosie and off we squeak down the corridor, her arm on mine, pausing each time she begins to feel the pressure within to adopt our practised positions: interlocking arms and hands the way ballroom dancers do just before the music starts.

The tawny owl has flown from the walls of the Antenatal Clinic along with the rest of the menagerie of animals. Other signs have replaced it: more tinselly adverts for baby photographers and a poster-paint globe cupped in a human hand. *Heal the World*, it instructs in rainbow letters tacked above. In a small room off to the side, Rosie sits up on a bed as a cardiotocograph asserts the vivacity of our baby's life via twitching needle and a jagged mountain range drawn along a roll of graph paper. Half an hour later and satisfied with the topography, the midwife beams at us: 'All good. You can probably go home now, to be honest.' She lifts up the half-watch clipped to her pocket. 'Baby could be hours away yet.' But we don't go home. Instead, we're moved up a flight of stairs to an empty observation bay. It is a peachy room in every sense, filled with clean beds, each with a pay-for-TV monitor suspended on an arm above it. The screensavers flick in sync between live news and adverts. After another contraction passes, we take the bed by the window. A horse chestnut folds its rusty leaves up against the glass. Beyond it, over a security fence, a few dog walkers and runners are crossing a thick, flat swathe of green edged with trees. It's a view and a half. You can see a long way from here, a long way back.

The maternity wing overlooks the Stray, the 200-acre horseshoe of common ground created in a famous gesture during the land enclosures in 1778. This wasn't to be 'common' in the sense understood before the acts were passed; it wasn't to sustain the landless or dispossessed poor. The Stray's limited grazing rights or 'Cattlegates' were strictly for those copyholders recognised as previously holding tenancy over the ground, and of the fifty cattlegates allocated, the devisees of baronet Sir Thomas Ingilby received twelve. Rather, the Stray was a gesture to appease the new and emerging class of landowners concerned that the privatisation of the mineral springs here would damage the wider area's reputation as a burgeoning English Spa. Responding to their petitions, the King (through his title 'Duchy of Lancaster') bequeathed them the land with the promise it would remain open common where all and sundry could enjoy free and unfettered access to the medicinal waters for ever, *without being subject to the payment of any acknowledgment whatsoever for the same, or liable to any action of trespass, or other suit, molestation, or disturbance whatsoever, in respect thereof*. Protected by law, this gift set in stone the future of the rural hamlets of High and Low Harrogate as well as the older township they fell under: Bilton-with-Harrogate. It created the environment whereby a grand resort could rise and flourish, attracting visitors to its wide avenues and tree-lined parks, its hostelries and assembly rooms. And, importantly, the unique concentration of springs set amid this rolling curve of public green. Those were different times. With the spa industry drying up for good in the 1930s, the Stray now finds itself tussling with the legacy – Harrogate is a tourist hotspot, residential jewel and conference venue that requires space more than it does springs. Zoom out and you see these 200 acres surrounded. Hemmed and threaded with roads, the Stray is now encroached from every direction, squeezed by the fine, imposing architecture of the 'old' town on one side and, on the other, the density of streets, houses, churches and schools that form its southerly suburbs.

I open the window as far as it will go. About three inches. With no one here to be offended, we eat our egg mayonnaise sandwiches and then conduct laps of the room, arm in arm, returning after each to rest and take in the vista of grass and trees. Dutifully, Rosie plugs in her iPod and closes her eyes. Through her earphones I can hear the faint sound of hypnobirthing affirmations. An American woman with a voice like silk: *The colour violet causes the mind to vibrate; all of nature is in tune with violet. Go deeper. You are a vehicle of nature. In tune with nature. Go even deeper. Now envisage yourself in a soft, green mist. Just as the earth springs forth life so too will your body . . .*

Sitting beside her on the bed, looking through the window, I can just about make out a small, hexagonal stone building, an elegant pump room constructed in the nine-teenth century over one of the Stray's famous iron or *chalybeate* wells. It is shut up now like an abandoned light-house in a sea of grass, an oddly ornate distraction for those idling in traffic. A toppled turret. Nothing more. A dead king's wishes about free water carry little weight these days.

Every spring in Harrogate is under lock and key. I'm not sure what that implies – perhaps they need to be for their own protection – but walking here sometimes after heavy rain I've found patches where those ancient iron and sulphur waters have leached back up through the boggy grass, pooling and puddling again in the Stray's dips and muddy corners. Birds flock to these mineral lagoons just as they have for millennia, before every stone, brick and human story was laid down here. To see that scene enduring among the queues of cars restores me in some small way.

Rosie's waters break at 3:30 p.m., halfway through another loop of the room. A different midwife, Jean – a short, kind-faced woman with glasses and grey hair dyed to blond – arrives with paper towels, checks her watch and makes a note. I'd no idea it was so late. Time has become an elastic concept outside the precise clockwork of the con-tractions, arriving now every four minutes, and for forty intense seconds. 'You're three centimetres dilated, too,' Jean says, peeling off a rubber glove. 'So I think we should move you to the labour ward.' She smiles at me. 'Let's call it a free upgrade.'

Jean fetches a wheelchair as I pack up our stuff. I have that same punched-gut tension as you feel the moments before stepping on a stage – that edginess. And it won't go. Rosie is pushed down the corridor past idle equipment and boxy incubation chambers, but I don't think she's taking any of it in. Her eyes are becoming more focused after each wave, as though she is staring inwards at something I can neither see nor hear. Past rows of drawers and a bright reception desk, we're shown into a room with a single, high, mechanical bed in its centre and a bathroom to one side. Jean and a nurse move automatically through the space, opening drawers and preparing equipment as they ask questions and strap a blood pressure monitor on Rosie's arm: *Do you want some water? Would you like to try a bath?* Polite as ever, Rosie answers each – *No, thank you* and *Yes, please* – then succumbs again, folding over the bed, interlocking hands with mine and releasing a long humming breath deep into the hospital sheet.

I've never heard of it before, but midwives rely on a kind of data record designed to draw order from the process of birth. It has a name – *Partogram*. It makes the details of labour, such as dilation, the baby's heart rate and the mother's vital signs visible and measurable so that any variations can be identified and investigated. Jean explains that she is starting one and then wires Rosie up accordingly. After handing her a little paper cup with pills in, I watch her write: *Paracetamol (1gm) and codeine phosphate (60mg) taken – declines further analgesia.*

Perhaps they've dimmed the lights or it's growing darker through the window but the monitor recording the baby's heart rate glows a bright orange and for a while I can't take my eyes off it. I stare at those digits, thinking of what's behind them: the life nosing its way out of its dark world, a fluid-lunged thing beginning to haul itself ashore through the breakers. The bath next door fills, cools and is run out again. We mean to reach it but after an hour of trying Rosie is back on the bed contorting with the tides inside. Together we wrestle the pull and the pain, locking our fingers, gripping, straining, stretching, and it dawns on me that I'm holding on to her as much as she is to me. Our heads pressed together, I'm whispering words of encouragement, knowing she's only hearing tones. And in between the crests of the contractions, Jean is talking her down, a trainer in our corner – *calm, calm, you're doing great* – as Rosie drags deep from the gas and air. And then it surges again and she cries against the inevitability of it, before channelling, breathing and bracing. Jean lifts her voice, coaxing the animal in her: 'OK, NOW PUSH THIS TIME. PUSH NOW. SHOUT IF YOU WANT TO. BUT PUSH. That's *brilliant.*' Except I'm seeing something else: each time Rosie pushes those orange digits on the monitor plummet. Jean glances at them too and makes hurried notes. *Decelerations*, they call these; the baby's heart rate is swooping low with every uterine contraction before swinging back up again as it subsides. It's difficult to watch; a necessity to bestowing life that seems to come close to ending it every time. *Come on. Come on* – I'm willing it – *come on, little thing.* Another hour or more with Rosie, head-down, rolling into the waves every two minutes and for a minute, but no further progress. It seems a futile torture and I can see she's tiring with every exertion; her face is red, exhausted and soaking with sweat. *But baby has to come now* – that's what Jean is saying. *Baby has to come.* By 6 p.m. it is concerning enough that she asks another midwife to send for the duty doctor.

A tall man in smart trousers, blue shirt and a white coat breezes in, smiles and introduces himself with a warm Nigerian lilt. He checks the Partogram and quietly confers with Jean. I hear words passed between them – *vertex, presenting, crowning.* Then he nods and asks in a sudden calm, loud tone, like a headmaster: 'Why don't we have this baby now, Rosemary? You know you can, don't you? You know you *will* do it. You just *have* to push now, Rosemary. When it comes...NOW.' Suddenly there are three of us willing her – *Yes! Yes! That's it* – as the current drags her down again. She pushes back against the pillows, eyes shut, biting her lips. And the noise she's making now is a soul-noise, an animal noise. I press my head into hers again, my arm around her shoulder, and I'm rocking and whispering all the

reassuring words I can think of. Any spell to conjure this life
from her and end the pain. I feel the sinews of her straining
neck and her iron strength and hear the juddering, bellowing
of deep, desperate lungs. She calls out again, a long howl,
which resolves into a sharp series of exhalations, each a
moan or a '*hooo*'. A commotion as Jean and the doctor
cheer and lean forwards and suddenly something else is with
us: a slick, bruise-coloured, blood-cowled form that Jean
attends to quickly but gently. She wipes it, clears its face and
then lays it shivering and unfolding on Rosie's chest. 'It's a
boy. A little boy,' says Rosie and then, 'Thomas. *Thomas.*'
And I'm face to face with a life that has fought its way to
this beginning, all the way from nothing, from eternity.
Thomas who, had things been different, might never have
been, but now squeaks in his mother's arms as some hitherto
unrealised part of my brain counts each of his strengthening
breaths. And with every one I'm becoming more lightheaded.
My heart is thumping in my chest. This brew of emotions
is strong; old waters are bubbling up through the grass.
Instincts. There are words they use in the books, words like
'wonder', but all are insufficient to relay the hugeness of the
shift, the acute brightness and sensitivity like your head's
been thrust through a door into a different room, as if it's
you that's just been born. And your mouth is asking 'Is he
OK?' once, twice, because you feel useless and you can't hear
properly and because you're too scared to do anything but
ask that dumb question and look. In fact, *that's* what the
books should tell you: that you can't stop looking and that,
from here on, there will be no end to your fascination. How
you are seeing in the present tense and differently, more like
the way a hawk sees: every lash, pore, patch of skin and
every shaking, stretching limb, every fingernail and toe; the
small exactness of the lips and those welded-shut eyes

scowling open and rolling towards the light. But you're not seeing with the fury of a predator identifying weakness; it's the attentiveness of adoration. You're thinking, *Careful! Be careful*, as though he is made of thin glass. The books should explain that this brings as much terror as euphoria and how you might not realise you have tears looping down to your jaw until the doctor tells you; how even the soft hospital blanket they place around his innocent little form can seem like a desecration of perfection.

Then, at some point while I'm distracted and staring, a different animal steals into the room. Jean has not ceased in her attendance of Rosie, her care necessary because the placenta didn't birth properly. The cord came away in her hand (*1820 hrs Valementous insertion* – the Partogram records), but I presume this must be a fairly common occurrence. No one seems too concerned. There are a couple of injections before the doctor is called back to perform another summoning and deftly removes the placenta. *Right*, I think, *that must be that.* 'You'll be fine now,' Jean confirms as she pops out of the door, 'so I'll leave you alone for a bit.' And then it's just the three of us wrapped up in each other, lulled into a beautiful calm until, weirdly, Rosie stops speaking to Thomas. Then altogether. Even making those soft, low mammal sounds has become too taxing for her. 'Are you OK?' She smiles, and then blinks wearily down at the boy. I kiss her forehead and notice how pale it has become. *Fatigue. The lights. Must be.* The garish strips have been flicked on above us. She closes her eyes and shuffles position, as if going to sleep. Her long, brown hair falls in twists across her face and arms; peat streams coursing through snowy moor. *She's too pale.* I frown. Then I hear the splash of water on stone. I take a step back and see blood spreading across the linoleum.

I must be shouting because Jean and a nurse run in exactly as another splatter spills sickeningly onto the floor. They both cry out 'Oh!' at the bright, scarlet pool. It is the movie blood of veins, arteries and haemorrhages. Panic hits me like a slap and I stroke Rosie's head: 'What is it, sweetheart? What's *wrong?*' But she won't – or can't – respond. Her arm goes limp and slides from Thomas, leaving him washed up on her breast. As the nurse slams the red alarm button over the bed, Jean scoops the baby up in a single movement and hands him to me. Then the doctor bursts in and suddenly I feel like I'm falling backwards or that the bed and its attendants are drifting away, the way a loosened boat slips from harbour. And now I see it and feel it, that wild animal that crept into the corner while my guard was down. I sense its size and shape; nature's other side, the chaotic antithesis of the hypno-birthing affirmations; this vicious twin of glorious creation and I'm thinking, *You cruel fucking thing*, to give and take in the same gesture, to open the heart and sharpen the senses, then do this. Leaning over her, the doctor is asking firmly and loudly: 'Rosemary? Rosemary?' And I want to yell at him – *Stop talking and DO something* – but they are trying, and I see that too. Drips are wheeled to the head of the bed; saline and bloods quickly plumbed into the back of her hands. There are more injections. Her blood pressure flashes on a monitor (88/50). Jean lays paper towels on the floor so nobody will slip. But I'm still slipping, further back to the window and to the darkness outside, struggling with the unfathomable weight of this baby staring up at me with its deep dark-blue eyes. 'It's OK,' I whisper with my lips touching his forehead. 'Everything's going to be all right.' But I'm not talking to Thomas; I'm haggling with that presence staring indifferently at me from the bed. *Please. Not this. Not this.* Of course, there's no arrangement you can make, no matter how hard you beg. It just glares back asking if I remember what being animal *really* means. And as Rosie lies there passed out, her blood darkening the paper towels, I realise I do. I'm afraid like I've never been before. The animal terror. 'Learn to fear,' advises J. A. Baker in his dark, apocalyptic book *The Peregrine*: 'To share fear is the greatest bond of all.' And I feel it now more deeply than I thought possible. *Fear*, the spark that ignites the flight of deer; that freezes the hare in its form; that fuels the owl's defence of its nest; that makes the fox caught in wire tear off its claws trying to escape. *Enough*, I say into Thomas's soft skin. *I've seen enough. Please stop.*

Gradually, and I mean *painfully* gradually, the injections start to work. The bleeding slows and then ceases altogether. After an hour Rosie stirs and starts to come around. Two more and her colour returns. Another and she's sitting up for the toast and sugary hot chocolate Jean has brought in on a tray. When I hand Thomas back to her, she's the one asking, 'Are you OK?' I'm the pale thing now, my arms cramped and trembling from holding the baby in the same position for the last four hours. Rosie, on the other hand, remembers nothing and is confused at where the time has gone. 'I think Dad might need a hug,' explains Jean and she comes over and puts her arms around me. After a moment I relent and sink into her hold. Wrapped in that human warmth, watching Rosie and Thomas burbling happily to each other again, I feel the fear withdrawing. Over Jean's shoulder, the animal has slipped from the room. There are other wards and beds to prowl; other hearts to bless and brutalise.

*

It's not far off two o'clock in the morning when I leave. Thomas and Rosie have been whisked off to a ward with the promise of more hot chocolate and buttered toast. But I'm not allowed to go with them. And they won't let me stay overnight, not even in a chair in reception. 'Go home,' the midwives say, laughing, 'we'll be watching them. Get some sleep.' But who are they trying to kid? There's too much stuff running round my head; too many revelations. I know where I'm going. I thank them all and ask them to pass on my heartfelt gratitude to the doctor busy bestowing calmness further down the corridor.

Outside it has turned deeply cold and the streets are deserted. Not another vehicle as I drive through the oily night, passing under the misted orbs of streetlamps along Skipton Road then right, down Bilton Lane. At the crossing point I stop, pull on my jacket from the boot and walk to the same fence I hunkered by on New Year's Eve. Wiping away the drop-in-temperature-tears with a sleeve, my eyes adjust. The silence thickens. Aside from a pylon showing as a deeper geometric darkness against the sky, the edge-land is an indistinct mass of blurry, coffee-black, all looming presence, distance and intimacy, exaggerated by Bilton's orange-washed roads and the few houselights still blazing over the fences. It is almost exactly like I found it those many months ago, only it doesn't feel strange any more.

When I first came to this spot I was seeking somewhere I might belong. I felt the urge to align myself with a place that, like me, seemed caught between states. Mapping this patch of ground has made it part of my life; we have blurred and planed together. It has altered my internal landscape even as I've watched it change. Perhaps this is a process that we all go through at some point, a kind of internal stock-take that occurs when confronted with the tectonic shifts in our existence, like moving away or impending fatherhood. There are times when we need to lose illusions and work out who we are, how we got here and where we're going. And now I realise how the outside world can inform our inside world. The common ground and edge-lands that surround our homes may not provide our food or fuel any more, but once unlocked, they can still sustain us, revealing the complex intermeshing between human and nature – showing us what we are, what we are not and how these two things are inseparable.

Despite the darkness, I know what lies beyond the fence. And, as I breathe, I pull this region close to me, drawing it into my lungs, conjuring visions of the precise shape of far hills, the lane and the woods, the hanging, grey silence of viaduct and gorge, the shuffling mice in the meadow, the dormant vetch seed in the soil, the starling shifting its hold on an electricity cable, the silent imprinting of a badger paw beside the holloway. I think of how beautifully telling it is that for all my time spent recording this edge-land's manifestation, of witnessing its histories and inhabitants coming to life, it still required that most human experience, a child being born, to feel the true sense and shape of being animal. And how, conversely, I feel all the more human for it.

Nearly twenty-four hours have elapsed since I stood in my kitchen waiting for the light to come, wrestling with the overbearing bleakness in this world. Sometimes it is impossible to come to terms with the things our species has done, and what it is capable of doing, but it can be easy to forget to hope too. And this is what I'm left with here and now. I write and circle a word in my notebook. HOPE. Even after everything there is hope because deep down people do care.

People *are* good. They take jobs that mean staying awake all night watching a ward of sleeping mothers and their newborn children, or they travel halfway around the world far away from their own families to care for the sick and dying on another continent. If someone stumbles on an escalator or falls in the street, the first instinct is not to steal their bag but to help them. I've witnessed that countless times and never before appreciated it for what it really is. To touch and reassure, to hasten over and bear-hug an emotional father in a maternity ward, kindness, compassion, the selflessness, the care, the heeding – these are natural states too. We need to fight to keep them alive and foremost, not surrender them to the other impulses our species carries within: selfishness, self-interest and one-upmanship.

As I was leaving the hospital I saw a face I recognised. A man in his thirties leaning with his back against the wall in a lower corridor, his eyes staring, brows jumping, as though he was running through a very serious conversation with himself. It was Danny, one of the dads-to-be who'd attended the same series of baby-care classes as us a few weeks back. After a long and difficult labour his wife had just birthed a boy. We talked for a few minutes and he looked a little shaky, frightened and tired, so I put my arm round his shoulder. He smiled and then, suddenly and forcibly, sobbed. 'Sorry,' he said immediately, 'I'm sorry.' There are times when the distance between us becomes noticeably less, when you recognise the humanity in others and feel the common thread that knots and ties us all together. In the same way the zoomed-out eyes of those first astronauts were gifted a unique perspective of this planet – the preciousness and precariousness of a small pale-blue dot in cold, sparse space – I feel the cogs that turn unstoppably under the surface; the connection we all share from living out our days together and, at the same time, the beauty and viciousness those days entail. I think of how we owe it to ourselves to make the best of it all during our short-lived stay. And I wonder whether, if we could hold on to such truths, the answer to that question – *what kind of world are we bringing you into?* – might yet be different.

Exhaustion and crashing emotions are catching up with me. I start back for the car, looking up at the sleeping town as I walk. The lights shivering and twinkling against the black remind me of those images you see of distant galaxies forming. I think of Thomas asleep in a crib with Rosie curled next to him, then of the endless potentiality within our grasp. And for a moment the world seems right.

'Oh and by the way, did you hear?' Rosie asks. An intense, excited, delirious week has already passed since we brought Thomas home. For the umpteenth time, changing his nappy is a two-person job, requiring fresh Babygro, blanket, cardigan, hat, even a sponging-down of our bedroom wall.

'Hear what?'

'It was on the news at lunchtime. They're calling off the badger cull.'

It's a funny feeling that follows, like just after you yawn or sneeze: a little rush, then a stolen second of stunned reflection.

Later, while Rosie feeds Thomas upstairs, I flick on the kitchen radio and listen to the Commons announcement being replayed in full. Turns out it is less an abandonment, more a stay of execution. At the dispatch box Owen Paterson sounds as defiant as ever, emphasising it is a 'postponement' until next summer, resolutely insisting that there is no shift in government policy. But there is a

shift, and I can feel it. I suspect he can too. Common
sense, public conscience and scientific reason have pre-
vailed and prevented – for now, at least. Who would have
thought it? Forcing a change through scrutiny, assessment
of evidence, resilience, determination and action; a
national show of compassion towards the non-human
world – sometimes even a little victory can go a long way
to restoring your faith. It lightens the horizon, even as the
days draw in and darken.

I am dreaming of the edge-land again, down in the midst
of Scots pines and the cold, scrub-scattered banks. I'm
walking by the river upstream, towards the viaduct,
following something. But *what*, I can't quite tell. A faint
shift in the silence is all. The wood is grey and black. The
river slips by. And I'm still going. Footstep after footstep.
And it's still going. Another slight *fizz* ahead, like a rustle,
a soft tread on leaves or a leg shifting under a blanket. I roll
my head on the pillow and, still dreaming, quicken my pace
to reach the trees that lean and stretch from the bank like
fingers pointing lazily northwards, over the Nidd. Another
movement, this one a snuffle, like a fox cub, and I realise
that I'm no longer looking through the trees but down into
a dark den or sett. And that whatever is in there is working
up to a scream. A pause and my mind reacts like a pilot
flame firing in a boiler, yanking me from sleep. My legs kick
out from under the duvet; my feet touch carpet and I push
and pivot my body, covering the few feet to the Moses
basket stationed at the end of our bed. I rub my eyes but
there's nothing to attend to. Thomas has settled himself.
No scream erupts. It was probably just wind. He is perfectly
happy, snug and swaddled with the fingers of his left hand

poking out of the top in a nonchalant wave. I smile and put my finger in his palm, letting him grip it. 'I'm here,' I whisper. 'Daddy's here'. A rookie move. He clamps down hard and I'm caught. I try to work my hand free but each time I do he stirs, whimpers and makes like he's about to shriek. I glance over at Rosie who is lost in the duvet and exhausted slumber, her arm tucked under the pillow. Anxious not to disturb her, I give up and stand in the dark watching Thomas sleep, wondering at the indescribable strength of the thing that's grasped hold of me. It takes a moment to straighten it out, but it's love, of course. It fills my head, my heart and this whole messy house.

Epilogue

THE NOTEBOOK

It has lain untouched on a low shelf by the door for well over a month. I've walked past countless times without incident, yet today I find the notebook has somehow secreted itself into one of the cavernous pockets of my warmest coat. Rummaging for a glove, my nail meets a familiar hard-backed cover. I curl my forefinger under the elastic band that keeps its pages closed, and pull it out with a frown of puzzlement. *Hello. How did you get there?* Between wrestling on boots and hauling the pushchair out through the narrow hallway, my hand must have instinctively reached out. I shouldn't be surprised, though; old habits die hard. Sometimes we know ourselves better than we think. This morning, for the first time, I'm walking Thomas down to the edge-land.

It's a creaking-cold day. Winter, unblinking, is staring down autumn, frightening it from its perch, chasing it from the woods with first frosts and sharp, fiery light. The sun shows willing but shines weakly from the east as I steer the pushchair down Bilton Lane, stopping every now and then to make sure the boy's all right. He's probably a bit over-dressed. On top of a thick cardigan I've bundled him up in a bright woolly hat, mittens and tucked two blankets under his muffler, but then again we've never walked this far from the house before and the air feels distinctly draughty, like a great sash window has been left open to the north. The sky has the same oily rainbow sheen as you find inside oyster shells and it stretches brilliantly behind roofs and telegraph wires, over hard, bare stubble fields, woods and the hills that range away unreachably between the bungalows and semis.

At the crossing point the two oaks either side of the road have been almost stripped completely. The sun paints branches yellow; beneath, the grass plays dead. Thrushes spill song and dogs bark. I'm grateful for the track they've laid over the old railway. The pushchair's wheels glide over the bitumen in a single, smooth, continuous note. Bikes whistle past us in both directions and light ripples the red, tiled rooftops of Tennyson Avenue. Reaching the meadow, I push us past the pylon and across country to the wood's edge, where I sit on a fallen pine branch. A few feet away a tiny goldcrest appears and begins working up and down a dead thistle, its Mohican of gold gleaming like a celestial streak. I assume Thomas must be fast asleep from the trundling, but when I turn the pushchair around I find his eyes are staring up in wonder, reflecting the trees and sky, taking everything in. Lucky thing. To have those days ahead when all the world is wide and bright, and all the world is all you can see.

I reach into my pocket and retrieve the notebook. It is a clutter of things, layered and brown like the earth, its cover wonderfully dirty. I undo its elastic fastening and flick back and forth through its pages, through the maps, scribbles of fox tracks, paragraphs of nineteenth-century railway research, drawings of the weir, a tramp's babblings, a folded-up brochure for Bilton Hall Nursing Home, dead leaves and squashed deer droppings, dried flower heads and stray Himalayan balsam seeds. There are wobbly, half-completed tables that note the timings of owl calls and the appearance of swifts above the fields; sketches of hares and the compass-point fields of vision matriarchal rabbits adopt when guarding their young. Here, pressed in a margin, is a mayfly's wing; here a found roach end and a squashed sprig of nettle. Here, a rough outline of a badger's print. Bound and folded together with it all, my thoughts and feelings and impressions. Brief moments in time. The extraordinary you can find in the ordinary. A changing life in this fleeting, wheeling world.

In many ways I wish I could have written a neater story but I suppose edge-lands, like our lives, are tangled stories that write themselves. And for all the pages we might fill, nothing comes close to a second of being here.

We sit quietly, the boy and me, until the goldcrest has gone. After a while I close the notebook, tuck it next to him and turn us both for home.

You showed me eyebright in the hedgerow,
Speedwell and travellers joy.
You showed me how to use my eyes
When I was just a boy;
And you taught me how to love a song
And all you knew of nature's ways:
The greatest gifts I have ever known,
And I use them every day.

Martin Simpson, 'Never Any Good'

Acknowledgements

The existence of this book is largely attributable to five extraordinary women who collectively birthed, nurtured and provided the space for all that is written here. Firstly my mother, Anne, whose encyclopedic knowledge of, and love for, nature and the outside world opened the door for me a long time ago. She has been a reliable source of information, a library where, mercifully, no fees are incurred, and an inspiration since she dragged my brother and I from our beds to go badger watching on Ilkley Moor. My wife, Rosie, never agreed to the details of our life being so frankly revealed in print and never complained when it became clear they might be. For so many reasons there simply would be no book without her. She is responsible for everything precious in my life, and to her I owe everything. Stephanie Ebdon had the vision to see potential in 150,000 words of scribbled field notes, odds-and-ends and ideas, and then the conviction and patience to find the right publisher. Her support has always been unflinching and hugely important. As has that of my agent, Jessica Woollard, who first believed in my writing many years ago and continues to be the most remarkably insightful, encouraging and brilliant ally and advocate. Last, but most certainly not least, my editor at Hutchinson, Sarah Rigby – a colleague and a good friend. When she undertook this work I doubt she had any idea of the journey we would be taking together, yet she has proved to be the kind of travelling companion every writer dreams of: visionary, loyal, inspiring, attentive and always on hand to console, cajole and put things in order. I am forever grateful to her for sharing with me her talent, her generosity of spirit and her contagious desire to kick at boundaries.

Thanks also to everyone else who worked on the book behind the scenes. At Hutchinson and Windmill: Jocasta Hamilton, Najma Finlay, Laura Deacon, Chris Turner and Matthew Ruddle. For the exquisite cover, Jason Smith, and for the gorgeous interior map of the edge-land, Emma Lopes. To Hannah Ferguson, Charlotte Bruton and all at The Marsh Agency, and the attentive Alison Tulett for keeping my words in check.

My love and thanks to my children, Thomas and Beatrice, my brother, Matthew, and my father, Maurice, for their knowledge, advice and kindness along the way; also to Karen, Natalie, Freya and Niamh and the rest of the wider, wilder Cowen family. For, variously, the long discussions, long walks, friendship, willingness to lend an ear or a spare room: Tim Jones and Zara Pearson, Alex and Eugenie Price, Charles Westropp and Lydia Sadler, Georgie Hoole and Mark Ballantyne, the Hoole family, Jordan and Alanna Frieda and family, Alex Corbet Burcher, James and Jess Westropp, Theo Cooper and Danielle Treanor, Stuart and Jill Smith and family. For their profound farming insight and knowledge: Andrew Sebire, Michael Flesher and the late Stanley Flesher; and for navigating me through the murky waters of Harrogate District's planning and housing development policies, Matthew Bagley. To all my friends for their love and support along the way, especially: Simon Skirrow and family, Rob Menzer and family, Sophie Smyth, Leo Critchley, Phil Westerman and Caroline Toogood, Will and Frankie Ridler, George and Alison Scott-Welsh, Lizzie and Danny Varian, Amy and Tom Holmes, James Yuill and Ruth Tapley, Fiona and Geoff Scholtes, Katie Sotheran and Ash Anderson, Xavier and Rachel Archbold, Paul Schofield and the much-missed Peter Westropp.

Various institutions have given me the space, literally and metaphorically, to write and research ideas. My thanks to: the *Telegraph* (especially Paul Davies), *Independent*, *Guardian* and all the staff at Harrogate and Ilkley libraries, especially Irene Todd. In recent years libraries have come under overwhelming financial strain, tasked with ever-greater areas of responsibility to the community. Writing this book has shown me how important such quiet spaces, repositories of local memory and gateways to new worlds, are – especially in a virtual age. These institutions and their staff deserve recognition, support and financing, as the day that we allow free and public centres of learning to degrade or disappear is a step down a dangerous road. I am hugely grateful to all those that sought to record Bilton's past and present before me, namely the local historians and naturalists of Bilton Historical Society, Bilton Conservation Group (especially the ever-inspiring Bill Williams and Keith Wilkinson), Knox Valley Residents' Association and Harrogate District Naturalists. History is imperfect and I have recorded fictions too real to not be true and truths that seem nothing short of fiction. The Bilton that appears here is mine; people's names have been changed in some cases, in others it was important that they had to remain the same. But all mistakes are my own fault and mostly intentional. My grateful thanks also to the staff and grounds team at Bilton Hall Nursing Home and the fantastic midwives, doctors and nurses of Harrogate District Hospital.

Profound thanks to Tim Dee for the words, the birds and the beers. He has been a great friend, inspiration and an invaluable resource during the thinking and writing of this book. To Martin Simpson for granting me permission to use the lyrics to 'Never Any Good' when I couldn't think of anything that said it better, and thanks also to the music of Mike Oldfield, John Martyn, Nic Jones, Nick Drake, George Harrison, Michael Chapman, Anne Briggs, Steeleye Span, Arcade Fire, Bob Dylan and Tom Waits. Lastly to all those others who have trodden literary paths into nature or place before me, and in whose footsteps I humbly follow: J.A. Baker, Richard Mabey, Roger Deakin, Robert Macfarlane, Kathleen Jamie, Helen Macdonald, Luke Jennings, Horatio Clare, John Clare, Richard Jeffries, Edward Thomas, Thomas Hardy, T.S. Eliot, J.L. Carr, Mark Cocker, W.A. Poucher, Richard Muir, Bill Williams, Marion Shoard, Nan Shepherd, Christina Hardyment, J.G. Ballard, W.R. Mitchell, Henry Williamson, W.G. Sebald, Patrick Barkham, George Monbiot, Tim Binding, Philip Hoare, Melissa Harrison, Simon Armitage, Olivia Laing, Nick Papadimitriou, Iain Sinclair, Will Self, Paul Farley, Michael Symmons Roberts, Michael McCarthy, Bill Griffiths, John Crace, Alan Bennett, Philip Larkin, Tony Harrison and Ted Hughes.

Notes and Selected Bibliography

Place of publication is London unless otherwise stated.

Epigraph

ix 'To know fully': Patrick Kavanagh, 'The Parish and the Universe', *Collected Pruse*, MacGibbon and Kee, 1964.

Prologue: New Year's Eve

Marion Shoard, 'Edgelands', as it appears in *Remaking the Landscape* (ed. Jennifer Jenkins), Profile Books, 2002.

Marion Shoard, *This Land is Our Land*, Gaia Books, 1997.

Oliver Rackham, *The History of the Countryside*, J.M. Dent, 1986.

Jean Clottes, *Chauvet Cave: The Art of Earliest Times*, University of Utah Press, 2003, translated by Paul G. Bahn, from *La Grotte Chauvet, l'art des origins*, Éditions du Seuil, Paris 2001.

David Lewis-Williams, *The Mind in the Cave*, Thames & Hudson, 2002.

11 'Becoming-animal': taken from Gilles Deleuze, 'The Body, the Meat and the Spirit: Becoming Animal' from Tracy Warr (ed.), *The Artist's Body*, translated by Liz Heron, Phaidon Press, 2000.

Simon Schama, *Landscape & Memory*, HarperCollins, 1995.

Crossing Point

Ely Hargove, *History of the Castle, Town and Forest of Knaresborough: with Harrogate, and Its Medicinal Waters*, York Edition, 1798.

Bernard Jennings, *A History of Harrogate and Knaresborough*, Advertiser Press, Huddersfield, 1970.

William Grainge, *History and Topography of Harrogate and the Forest of Knaresborough*, Smith & Co., 1871.

Bill Williams, *Bilton Through the Ages,* Bill Williams, Harrogate, 1985.

Walter J. Kaye, *Records of Harrogate*, F.J. Walker, Leeds, 1922.

Phyllis Hembry, *The English Spa, 1560–1815,* The Athlone Press, 1990.

G. Body, *PSL Field Guides – Railways of the Eastern Region*, Volume 2, Patrick Stephens Ltd, Wellingborough, 1988.

Walter Weaver Tomlinson, *The North Eastern Railway, its Rise and Development*, Longmans, Green and Co., 1914.

On the development of Harrogate's railway lines, see: http:// lostrailwayswestyorkshire.co.uk/Leeds%20Harrogate.htm.

Malcolm Neesam, *Harrogate Great Chronicle, 1332–1841*, Carnegie Publishing, Lancaster, 2005.

H.G. Lloyd, *The Red Fox*, Batsford, 1980.

Martin Wallen, *Fox*, Reaktion Books, 2006.

Kenneth Varty, *Reynard The Fox*, Leicester University Press, Leicester, 1967.

Laura Spinney, 'The Lost World', *Nature*, 454, 2008.

Derek Yalden, *The History of British Mammals*, T. & A.D. Poyser, 1999.

Richard Jefferies, *The Gamekeeper at Home* and *The Amateur Poacher*, Oxford University Press, Oxford, 1960.

Hope P. Werness, *The Continuum Encyclopedia of Animal Symbolism in Art*, Continuum, 2003.

C.G. Jung, *The Archetypes and the Collective Unconscious*, Routledge (2nd edition), 1991.

C.G. Jung, *Psychological Types*, translated by R.F.C. Hull as in *The Collected Works of C.G. Jung*, Bollingen Ser 20, Volume 6, Princeton University Press, Princeton, 1971.

Laurie Milner, *Leeds Pals: History of the 15th (Service) Battalion (1st Leeds) the Prince of Wales' Own (West Yorkshire Regiment), 1914–18*, Wharncliffe Books, Barnsley, 1993.

Chris Mead, *Robins*, Whittet Books Ltd, 1989.

Ultrasound

Bertel Brunn (with illustrations by Arthur Singer), *Birds of Europe*, McGraw-Hill, 1971.

John R. Mather, *Birds of the Harrogate District*, Harrogate and District Naturalists' Society, Harrogate, 2001.

Chris Mead, *Owls*, Whittet Books Ltd, Stansted, 1994.

Leanne Thomas (with illustrations by Chris Shields), *Guide to British Owls and Owl Pellets,* Field Studies Council, 2008.

Colin Harrison, *A Field Guide to Nests, Eggs, Nestlings of British and European Birds*, Viking Press, 1982.

Mark Cocker and Richard Mabey, *Birds Britannica*, Chatto & Windus, 2005.

Mark Cocker, *Birds and People*, Jonathan Cape, 2013.

Paul Sterry, *Collins Complete Guide to British Birds*, Collins, 2008.

On tawny owl calls and mating, see: http://godsownclay.com/ TawnyOwls/tawnyowlsentrypa.html.

Philip Stewart Robinson, *Birds of the Wave and Woodland*, Hard Press, 2013.

Krystyna Weinstein, *The Owl in Art, Myth, and Legend*, Grange Books, 1990.

Virginia C. Holmgren, *Owls in Folklore and Natural History*, Capra, Santa Barbara, 1989.

Gertrud Benker, *While Man and Nature Sleep: Owls are Cultural Symbols of Dark Mystery and Good Fortune*, The World & I Online (Kindle edition), 1993.

71 'She is a bird indeed . . . but conceals her shame in the darkness; and by all the birds she is expelled entirely from the sky': Ovid, *The Metamorphoses* (FABLE IX), translated by Henry T. Riley, Digireads.com, 2009.

71 'He is the very monster of the night . . . if he be seen, it is not for good, but prognosticates some fearful misfortune': Pliny (The Elder), *Natural History: A Selection*, translated by John F. Healey, Penguin, 1991.

71 'It was the owl that shrieked, the fatal bellman, which gives the stern'st goodnight': William Shakespeare, *Macbeth*, 2.ii, 4–5.

D.W. Snow and C.M. Perrins, *The Birds of the Western Palearctic* (concise edition), Oxford University Press, Oxford, 1998.

Robert Burton, *Bird Behaviour*, Granada Publishing, 1985.

Karel H. Voous, *Owls of the Northern Hemisphere*, Collins, 1988.

81 'All of the night was quite barred out except . . .': Edward Thomas, 'The Owl', as it appears in *Annotated, Collected Poems*, edited by Edna Longley, Bloodaxe Books, Hexham, 2008.

The Union of Opposites

83 'A change in the weather': Marcel Proust, *In Search Of Lost Time, Volume 3: The Guermantes Way*, translated by C.K. Scott Moncrieff, D.J. Enright and Terence Kilmartin, Vintage Classics, 1996.

George Ewart Evans, David Thomson, *The Leaping Hare*, Country Book Club (reprint edition), 1974.

87, 92, 96, 98 'The man who encounters the hare': the sections of folk terms for the hare throughout this chapter are taken from the Middle English poem 'The Names of the Hare in English', which appears in many forms. This is the version transcribed by Ewart Evans and David Thomson in *The Leaping Hare*, Country Book Club (reprint edition), 1974.

John Layard, *The Lady of the Hare*, Shambhala, Boston, 1988.

Simon Carnell, *Hare*, Reaktion Books, 2010.

Christine Gregory, *Brown Hares in Derbyshire: The Story of One of the Peak District's Most Enigmatic Mammals*, Vertebrate Publishing, Sheffield, 2012.

Heather McDougall, 'The pagan roots of Easter', *Guardian*, 3 April 2010.

Adrian Bott, 'The modern myth of the Easter bunny', *Guardian*, 23 April 2011.

90 'The hare springs up the hero in many flood myths': this telling of the hare's role in saving Noah's Ark is explained by Alan Dundes and appears in his *The Flood Myth*, University of California Press, Oakland, 1988.

91 'For one month it becomes male, and the other female': this example is taken from *Ancient Laws and Institutes of Wales: laws supposed to be enacted by Howel the Good*, an insightful and revealing repository printed in 1831.

Jeremy Rifkin, *Biosphere Politics: A New Consciousness for a New Century*, Crown Publications, New York, 1991.

The Anglo-Saxon Chronicle, translated by George Norman Garmonsway, Everyman's Library, 624, 1953.

Kevin Cahill, *Who Owns Britain*, Canongate, Edinburgh, 2001.

Simon Fairlie, 'A Short History of Enclosure in Britain' as it appeared in *The Land*, Issue 7, 2009.

Colin Ward, *Cotters and Squatters: The Hidden History of Housing*, Five Leaves Publication, 2002.

Claire Joy, 'The roots of our rootlessness – a history of enclosures', as it appears on http:/www.reclaimthefields.org.uk, accessed 2 April 2012.

George Monbiot, 'The Tragedy of Enclosure', in *Scientific American*, 1994.

Joan Thirsk, 'The Common Fields' as it appeared in *Past and Present*, 29, 1964.

Giselle Byrnes, *Boundary Markers*, Bridget Williams Books, New Zealand, 2002.

C.S. and C.S. Orwin, *The Open Fields*, Clarendon Press, Oxford, 1938.

Thomas More, *Utopia*, Everyman, 1994.

'The Harrogate Improvement Act being an act for improving certain parts of the townships of Bilton with Harrogate, and Pannal, called High and Low Harrogate, for protecting the Mineral Springs and for other purposes therein mentioned; (Verbatim from the Parliamentary Copy) Prepared by the Solicitors for Obtaining the Act', Pickersgill Palliser, Harrogate, 1845.

James Manby Gully, *Water Cure in Chronic Disease*, Simpkin, Marshall & Co., 1884.

Gillian Thomas, 'Lime and Coal in Bilton' as it appeared in St John Bilton's Parish Magazine, date unrecorded.

Bernard Jennings, *A History of the Wells and Springs of Harrogate*, Harrogate Corporation Department of Conference and Resort Services, Harrogate, 1974.

Adam Hunter, *A Treatise on the Mineral Waters of Harrogate and its Vicinity*, Longman and Co., 1830.

Neville Wood (ed.), *Health resorts of the British Islands*, University of London Press: Hodder & Stoughton, 1912.

Andrew Scott Myrtle and James Aitken Myrtle, *The Harrogate Mineral Waters and Chronic Diseases, with Cases*, R. Ackrill, Harrogate, 1867.

John B.L. McKendrick, 'The Value of Sulphur Baths in Rheumatism', a Thesis for Degree of M.D at Glasgow University, 1934.

Robert Mortimer Glover, *On Mineral Waters: Their Physical and Medicinal Properties*, Henry Renshaw, 1857.

Malcolm Neesam, *Harrogate Great Chronicle, 1332–1841*, Carnegie Publishing, Lancaster, 2005.

William Grainge, *History and Topography of Harrogate and the Forest of Knaresborough*, Smith & Co., 1871.

Bill Williams, *Bilton Through the Ages,* Bill Williams, Harrogate, 1985.

Walter J. Kaye, *Records of Harrogate*, F. J. Walker, Leeds, 1922.

Richard Muir, *Shell Guide to Reading the Landscape*, Michael Joseph Ltd, 1981.

137 'There are voices today that call for "re-wilding"': I'm grateful to Tim Dee for our numerous conversations on modern forms of enclosure and the challenges of nature conservation and mediated experiences of landscape. He also touches on these subjects in his wonderful *Four Fields*, Jonathan Cape, 2013.

DNA

Richard Prior, *The Roe Deer: Conservation of a Native Species*, Swan Hill Press, Shrewsbury, 1995.

Henry Tegner, *The Roe Deer – Their History, Habits and Pursuit*, Tideline Publications, Rhyl, 1981

Emma Griffin, *Blood Sports: Hunting in Britain since 1066*, Yale University Press, Yale, 2008.

Ruth A. Johnston, *All Things Medieval: An Encyclopedia of the Medieval World, Volume 1*, ABC-CLIO LLC, 2011.

Richard Almond, *Medieval Hunting*, The History Press, 2011.

John Cummins, *The Hound and the Hawk: Art of Medieval Hunting*, Weidenfeld & Nicolson, 1988.

Edward of Norwich, edited by William A. and F.N. Baillie-Grohman, *The Master of Game*, The History Press (reprint edition), 2011.

Mike Brough, *History & Hikes of the Ancient Royal Hunting Forest of Knaresborough*, Colin Michael Brough, Harrogate, 2013.

Bernard Jennings (ed.), *A History of Nidderdale*, Pateley Bridge Tutorial Class; Advertiser Press Limited, Huddersfield, 1983.

Early Yorkshire Charters Volumes I and IX, based on manuscripts of the late William Farrer, Charles Travis Clay CB, FBA (ed.), Yorkshire Archaeological Society Record Series, 1952.

Arnold Kellett, *Historic Knaresborough*, Smith Settle Ltd, Otley, 1991.

Arnold Kellett, *The Knaresborough Story*, Lofthouse Publications, Pontefract, 1990.

Nicky Milner, Barry Taylor, Chantal Conneller and Tim Schadla-Hall, *Star Carr: Life in Britain After the Ice Age*, Council for British Archaeology, 2013.

J.G.D. Clark, *Excavations at Star Carr*, Cambridge University Prress, Cambridge, 1954.

T. Clare, 'Before the first woodland clearings', *British Archaeology 8*, http://www.britarch.ac.uk/ba/ba8/BA8FEAT.HTML, 1995.

R. Chatterton, 'Star Carr reanalysed', in J. Moore and L. Bevan (eds.) *Peopling the Mesolithic in a Northern Environment*, Archaeopress, Oxford, British Archaeological Reports International Series 1157, 2003.

C.J. Conneller, 'Star Carr recontextualised', in J. Moore and L. Bevan (eds), *Peopling the Mesolithic in a Northern Environment*, Archaeopress, Oxford, British Archaeological Reports International Series 955, 2003.

Brian G. Dias & Kerry J. Ressler, 'Parental olfactory experience influences behavior and neural structure in subsequent generations', *Nature Neuroscience 17*, 2013.

Nessa Carey, *The Epigenetics Revolution: How Modern Biology is Rewriting Our Understanding of Genetics, Disease and Inheritance*, Icon Books Ltd, 2012.

Benjamin Elliott, 'Antlerworking practices in Mesolithic Britain', PhD thesis, University of York, Department of Archaeology, 2012.

One Day

E. Percival and H. Whitehead, *Observations on the Biology of the Mayfly, Ephemera Danica, Mull,* reprint from the Leeds Philosophical Society, Leeds, 1926.

Malcolm Greenhalgh, *The Mayfly and the Trout*, The Medlar Press, Ellesmere, 2011.

Paul Giller and Bjorn Malmqvist, *The Biology of Streams and Rivers*, Oxford University Press, Oxford, 1998.

Gary Lafontaine, *The Challenge of the Trout*, Mountain Press, Missoula, 1983.

The Turning Time

Chris Kightley, Steve Madge and Dave Nurney, *The Pocket Guide to* the Birds of Britain and North-West Europe, Pica Press, 1998.

R.S.R. Fitter, Collins Pocket Guide to British Birds, Collins, 1966.

Mark Cocker and Richard Mabey, Birds Britannica, Chatto & Windus, 2005.

David Lack, Swifts in a Tower, Methuen, 1956.

Derek Bromhall, Devil Birds: The Life of the Swift, Hutchinson, 1980.

Andrew Lack and Roy Overall, The Museum Swifts, Oxford University Museum of Natural History, Oxford, 2002.

Graham Appleton, 'Swifts start to share their secrets', British Trust for Ornithology (BTO) News, May–June 2012.

Håkan Karlsson, Per Henningsson, Johan Bäckman, Anders Hedenström and Thomas Alerstam, 'Compensation for wind drift by migrating swifts', Animal Behaviour, Volume 80, Issue 3, September 2010.

Johan Bäckman and Thomas Alerstam, 'Harmonic oscillatory orientation relative to the wind in nocturnal roosting flights of the swift Apus apus', Journal of Experimental Biology, 11 December 2001.

Elizabeth Day, 'Revealed: how the swift keeps to its course at 10,000ft – even as it sleeps', Telegraph, 14 March 2004.

Robert Furness, J. D. D. Greenwood, Birds as Monitors of Environmental Change, Springer (soft-cover reprint of 1st edition, 1993), 2013.

Bob Hume, 'Learning About Birds: Swift', Birds Magazine (RSPB), as it appears on http://www.stocklinch.org.uk/Swift.htm, accessed 29 July 2012.

Peter Berthold, Hans-Günther Bauer and Valerie Westhead, Bird Migration: A General Survey, Oxford University Press, Oxford, 2001.

S. Åkesson, R. Klaassen, J. Holmgren, J. W. Fox and A. Hedenström, 'Migration routes and strategies in a highly aerial migrant, the common swift Apus apus, revealed by light-level geolocators', PLoS One, http://www.ncbi.nlm.nih.gov/pubmed/22815968, 2012.

189 'They've made it again': Ted Hughes, 'Swifts', as featured in *The Poetry of Birds*, Viking, 2009, edited by Simon Armitage and Tim Dee.

206 'Naturalist Richard Mabey wrote of being struck by an unavoidable allegory': the article in question is from 2009 and appears in his collection *A Brush With Nature: 25 years of personal reflections on nature*, BBC Books, 2010.

W. Wiltschko, U. Munro, H. Ford and R. Wiltschko, 'Bird navigation: what type of information does the magnetite-based receptor provide?', *Proceedings of the Royal Society*, B 273 (1603), http://www.ncbi.nlm.nih.gov/pmc/articles/PMC1664630/, 22 November 2006.

Mark E. Deutschlander, John B. Phillips and S. Chris Borland, 'The Case for Light-Dependent Magnetic Orientation in Animals', *Journal of Experimental Biology*, 22 March 1999.

For more on the common swift, its diet, migrations, habits and habitats, see (and support) the fantastic: http://www.swift-conservation.org.

Metamorphoses

217 'Now I am ready to tell how bodies are changed': Ted Hughes, *Tales from Ovid: 24 Passages from the Metamorphoses*, Faber & Faber, 2002.

R.H Deaton, *Small Rodents of the Harrogate District*, Harrogate and District Naturalists' Society, Harrogate, 1988.

M. Barnham and G.T. Foggitt, *Butterflies in the Harrogate District*, Harrogate and District Naturalists' Society, Harrogate, 1987.

Joseph Moucha, *A familiar colour guide to familiar Butterflies, Caterpillars and Chrysalides*, Octopus, 1978.

Ferris Jabr, 'How Does a Caterpillar Turn into a Butterfly?', in *Scientific American*, 10 August 2012.

Editorial, M. S. Warren et al., 'Rapid responses of British butterflies to opposing forces of climate and habitat change', *Nature*, 414, 65–9, 1 November 2001.

234 'A biting east wind brought the temperature down to -1 °C':

Newsletter No. 17 of the Bilton Conservation Group, Winter 1985/6, 'The Wildlife of the Nidd Gorge in Winter' (Secretary: Keith Wilkinson).

Newsletter of the Bilton Conservation Group dated 'Spring 2007'.

Study by Cobham Resource Consultants for the Council of the Borough of Harrogate, 'Nidd Gorge and Bilton Fields: Proposals for the Management of the Countryside on the northern and eastern fringes of Harrogate', March 1983.

Peter Barnes (ed.), 'Bilton with Harrogate – Forest, Farms and Families', Report of a Community Archaeological Project by Bilton Historical Society, 2008.

Christopher Brickell (ed.-in-chief), *The Royal Horticultural Society Encyclopedia of Gardening*, Dorling Kindersley Ltd, 1992.

Richard Mabey, *Flora Britannica*, Chatto & Windus/Sinclair Stevenson, 1996.

David D. Stuart, *Buddlejas (Royal Horticultural Society Plant Collector Guide)*, Timber Press, 2006.

For more information on SUSTRANS, the sustainable transport charity, see: http://www.sustrans.org.uk.

For more information on butterflies as an indicator species, including reports, see the United Kingdom Butterfly Monitoring Scheme: http://www.ukbms.org/indicators.aspx.

Last Orders

Various local history sources and websites proved invaluable in the research of the fate of the 'Leeds Pals' during the First World War. Few records prove as evocative however, as the words of Private A.V. Pearson, speaking about his battalion after the war: 'The name of Serre and the date of 1st July is engraved deep in our hearts, along with the faces of our *Pals*, a grand crowd of chaps. We were two years in the making and ten minutes in the destroying.' An overview of the battalion's formation can be found on the Wartime Memories Project website (archived by the British Library), which also sheds light on the ferocity faced by its soldiers on their first engagement in the war: 'The 15th Battalion, West Yorkshire Regiment (1st Leeds Pals) was raised in

Leeds in September 1914 by the Lord Mayor and City. After training locally, the Battalion joined the 93rd Brigade, 31st Division, in May 1915 and moved to South Camp, Ripon and later to Hurdcott Camp near Salisbury. In December that year they set sail for Alexandria in Egypt to defend the Suez Canal before, in March 1916, the entire 31st Division left Port Said aboard *HMT Briton* bound for Marseilles in France, a journey that lasted five days. They travelled by train to Pont Remy, a few miles south-east of Abbeville, and marched to Bertrancourt arriving on 29 March 1916. Their first taste of action was at Serre on the first day of the Somme where they suffered heavy casualties as the battle was launched.' For more information see: http://www.wartimememoriesproject.com/greatwar/allied/westyorkshireregiment15.php and http://www.leeds-pals.com.

For a comprehensive breakdown of the battle of Serre on 1 July 1916, see: http://www.ww1battlefields.co.uk/somme/serre.html.

Recorded losses for the battalion, sustained in the few minutes after Zero, were 24 officers and 504 other ranks, of which 15 officers and 233 other ranks were killed. Despite the Commander-in-Chief, Douglas Haig, commenting negatively on the performance of VIII Corps on the 1 July 1916 in his diary – Gary Sheffield and John Bourne (ed.), *Douglas Haig: War Diaries & Letters 1914–1918*, Weidenfeld & Nicolson, 2005 – the Official British History recognises the efforts made by the men, many of who had never been in action before. In Brigadier-General Sir James E. Edmonds, *Military Operations France and Belgium 1916, Sir Douglas Haig's Command to the 1st July: Battle of the Somme: Volume I*, 1932, it states: 'There was no wavering or attempting to come back, the men fell in their ranks, mostly before the first hundred yards of No Mans Land had been crossed. The magnificent gallantry, discipline and determination displayed by all ranks of this North Country division were of no avail against the concentrated fire-effect of the enemy's unshaken infantry and artillery.'

Laurie Milner, *Leeds Pals: History of the 15th (Service) Battalion (1st Leeds) the Prince of Wales' Own (West Yorkshire Regiment), 1914–18*, Wharncliffe Books, Barnsley, 1993.

Major and Mrs Holt, *Battlefield Guide to the Somme*, Pen and Sword Books Ltd, Barnsley, 2008.

Chris McCarthy, *The Somme: The Day-by-Day Account*, Caxton Editions, 2000.

Martin Middlebrook, *The First Day on the Somme*, Penguin Books, 1984.

Martin and Mary Middlebrook, *The Middlebrook Guide to the Somme Battlefields: A Comprehensive Coverage from Crecy to the World Wars*, Pen and Sword Books Ltd, 2007.

Paul Reed, *Walking the Somme*, Pen and Sword Books Ltd, Barnsley, 1997.

Dylan Trigg, 'The Memory of Place: a Phenomenology of the Uncanny', as it appears on academia.edu: http://www.academia.edu/355785/The_Memory_of_Place_a_Phenomenology_of_the_Uncanny.

Prepared by order of W. Patrick, Esq., 'The Particulars and Plan of Bilton Hall Estate, to be offered for sale by auction by Renton & Renton, Wednesday the 4th day of June, 1924', accessed at Harrogate Library, 2012.

'North Harrogate Residents Consultative Group', pamphlet detailing proposals for housing and relief road to Skipton Road through Knox, Bachelor Gardens, Bilton, Starbeck, 1985. Under 'Fact One': 'The six miles of road is projected as a three-lane highway. It is more than an idea, it is an imminent reality'; under 'Fact Two': 'The intended road will facilitate a huge housing and industrial venture, expanding Harrogate beyond recognition', accessed at Harrogate Library, 2012.

Tony Cheal, *Bilton Residents 1882–1899*, listing document produced for the Bilton Historical Society, Harrogate, May 1996.

Jack Kerouac, *On the Road*, Penguin Modern Classics, 2000.

Revelations

Michael Clark, *Badgers*, Whittet Books Ltd, 1988.

Fiona Harvey, 'Badger culling is ineffective, says architect of 10-year trial', *Guardian*, 11 July 2011.

Patrick Barkham, 'Slaughtering badgers is not the answer to bovine TB', *Guardian*, 15 December 2011.

Damian Carrington and Jamie Doward, 'Badger cull "mindless", say scientists', *Observer*, 13 October 2012.

Patrick Barkham, 'Do we have to shoot the badgers?', *Guardian*, 6 August 2011.

Richard Black, 'Badger cull "may worsen problem"', BBC News, Science/Environment, http://www.bbc.co.uk/news/science-environment-11303939, 15 September 2010.

Angela Cassidy, 'Vermin, Victims and Disease: UK Framings of Badgers In and Beyond the Bovine TB Controversy', *Sociologia Ruralis*, Volume 52, Issue 2, April 2012.

289 On Owen Paterson tabling 600 questions regarding badger control in opposition, see Hansard, Common Debates, Column 1062, http://www.publications.parliament.uk/pa/cm201213/cmhansrd/cm121025/debtext/121025-0001.htm, 25 October 2012.

Damian Carrington, 'Owen Paterson: true blue countryman putting wind up green campaigners', *Guardian*, 11 October 2012.

Damian Carrington, 'Owen Paterson's climate change problem: cock-up or conspiracy?', *Guardian*, 7 September 2012.

290 'In a body blow to the new environment minister, thirty eminent British scientists': Professor Sir Patrick Bateson et al., 'Culling badgers could increase the problem of TB in cattle', open letter, *Observer*, 14 October 2012: http://www.theguardian.com/theobserver/2012/oct/14/letters-observer.

'In for the cull: A government that asks for independent scientific advice had best be ready to take it': Editorial, *Nature*, 450 (7166), 1–2, 1 November 2007.

Stephen P. Carter, Mark A. Chambers, Stephen P. Rushton, Mark D.F. Shirley, Pia Schuchert, Stéphane Pietravalle, Alistair Murray, Fiona Rogers, George Gettinby, Graham C. Smith, Richard J. Delahay, R. Glyn Hewinson and Robbie A. McDonald, 'BCG Vaccination Reduces Risk of Tuberculosis Infection in Vaccinated Badgers and Unvaccinated Badger Cubs', *PLoS ONE*, http://www.ncbi.nlm.nih.gov/pmc/articles/PMC3521029, 2012.

'Bovine TB: The Scientific Evidence – A Science Base for a Sustainable Policy to Control TB in Cattle – An Epidemiological Investigation into Bovine Tuberculosis', presented by Professor John Bourne (chairman of Independent Scientific Group on Cattle TB) to the Secretary of State for Environment, Food and Rural Affairs, The Rt Hon David Miliband, MP, June 2007. This full version of the report, including John Bourne's covering letter, appears on the Team Badger website and may be read in full at https://www.teambadger.org/pdf/final_report.pdf.

W.J. Stokoe (compiled by), Maurice Burton, D.Sc. (revised by), *The Observer's Book of Wild Animals of the British Isles*, Frederick Warne & Co., 1958.

Gareth Enticott, 'The Spaces of Biosecurity: Prescribing and Negotiating Solutions to Bovine Tuberculosis', *Environment and Planning A*, Volume 40, No. 7, 2008.

292 'I know I'm not alone in wondering': the insight and research of the *Guardian* between 2011 and 2012 were important in my investigations into the efficacy of badger culling. I am particularly grateful to Patrick Barkham and Damian Carrington and recommend Patrick's subsequent book *Badgerlands*, Granta, 2013, as a definitive work on the period.

Margaret Blount, *Animal Land*, Hutchinson, 1974.

Timothy Roper, *Badger*, HarperCollins, 2010.

Hannah Kuchler, 'Coalition clashes on wind farm policy', *Financial Times*, 31 October 2012.

Kiran Stacey, 'Allies warn Cameron of move to right', *Financial Times*, 5 September 2012.

Intergovernmental Panel on Climate Change (IPCC), 'Climate Change 2007: Synthesis Report', as it appears on https://www.ipcc.ch/publications_and_data/ar4/syr/en/spms2.html, 2007.

Intergovernmental Panel on Climate Change (IPCC), 'Climate Change 2007: The Physical Science Basis: Summary for Policymakers', as it appears on http://www.slvwd.com/agendas/Full/2007/06-07-07/Item%2010b.pdf, 2007.

For explanations of climate-change science and rebuttal of misinformation, see: http://www.skepticalscience.com

Camille Parmesan and Gary Yohe, 'A globally coherent fingerprint of climate change impacts across natural systems', *Nature*, 421, 37–42, 2 January 2003.

Cynthia Rosenzweig et al., 'Attributing physical and biological impacts to anthropogenic climate change', *Nature*, 453, 353–7, 15 May 2008.

George Monbiot, 'Bonfire of Promises', *Guardian*, 28 May 2012, and 'Shale Fail', published on the *Guardian* website, 31 August 2011, also featured on his website: www.monbiot.com.

Graham Readfearn, Leo Hickman and Rupert Neate, 'Michael Hintze revealed as funder of Lord Lawson's climate thinktank', *Guardian*, 27 March 2012: http://www.theguardian.com/environment/2012/mar/27/tory-donor-climate-sceptic-thinktank.

Steve Connor, 'How the "Kochtopus" stifled green debate: behind the climate "countermovement" are two billionaire brothers', *Independent*, 24 January 2013.

Ian Johnstone, 'Nigel Lawson's climate-change denial charity "intimidated" environmental expert', *Independent*, 11 May 2014.

Bob Ward, 'When climate change denial is promoted in mainstream news', *New Statesman*, 20 August 2014: http://www.newstatesman.com/politics/2014/08/when-climate-change-denial-promoted-mainstream-news.

'Global Warming's Denier Elite', Editorial, *Rolling Stone*, 12 September 2013.

'Climate Denial Machine Fuelled by Big Oil…', Editorial, as it appears on ecowatch.com, 29 July 2014.

For more information on the campaign against climate change, see: http://www.campaigncc.org

George Monbiot, *Heat*, Penguin, 2006.

Epigraph

'You showed me': Martin Simpson, 'Never Any Good', *Prodigal Son*, Topic Records, 2007.